SpringerBriefs in Electrical and Computer Engineering

Series editors

Woon-Seng Gan, School of Electrical and Electronic Engineering, Nanyang Technological University, Singapore, Singapore

C.-C. Jay Kuo, University of Southern California, Los Angeles, CA, USA

Thomas Fang Zheng, Research Institute of Information Technology, Tsinghua University, Beijing, China

Mauro Barni, Department of Information Engineering and Mathematics, University of Siena, Siena, Italy

SpringerBriefs present concise summaries of cutting-edge research and practical applications across a wide spectrum of fields. Featuring compact volumes of 50 to 125 pages, the series covers a range of content from professional to academic. Typical topics might include: timely report of state-of-the art analytical techniques, a bridge between new research results, as published in journal articles, and a contextual literature review, a snapshot of a hot or emerging topic, an in-depth case study or clinical example and a presentation of core concepts that students must understand in order to make independent contributions.

More information about this series at http://www.springer.com/series/10059

Waleed Ejaz • Alagan Anpalagan

Internet of Things for Smart Cities

Technologies, Big Data and Security

Waleed Ejaz
Thompson Rivers University
Kamloops, BC, Canada

Alagan Anpalagan
Ryerson University
Toronto, ON, Canada

ISSN 2191-8112   ISSN 2191-8120   (electronic)
SpringerBriefs in Electrical and Computer Engineering
ISBN 978-3-319-95036-5   ISBN 978-3-319-95037-2   (eBook)
https://doi.org/10.1007/978-3-319-95037-2

Library of Congress Control Number: 2018954057

This Springer imprint is published by the registered company Springer Nature Switzerland AG
The registered company address is: Gewerbestrasse 11, 6330 Cham, Switzerland

This book is dedicated to our families.

Foreword

Internet of Things (IoT) has now become a key enabling technology that spans multiple technology domains from data sensing and processing to networking and data analytics. IoT is used in many applications ranging from home security and factory automation to healthcare delivery to autonomous driving. In this book, the authors provide an essential overview of IoT for smart cities and key challenges associated with it, then cover communication technologies and protocols for IoT in smart cities. The coverage includes big data impacts on IoT operations, in terms of processing, storage, and analytics; security and privacy issues; and challenges of IoT for smart cities. IoT-based charging solution for electric vehicles is demonstrated as a practical application in smart cities. As such, it will be a good reference resource for graduate students, researchers, and industry practitioners working in IoT applications for smart city.

University of Idaho   Prof. Mohsen Guizani
Moscow, ID, USA   Ph.D., FIEEE
September 2018

Preface

The concept of the smart city was introduced as the potential solution to the challenges created by urbanization with complex and costly operations. The envisioned goal of smart city is to be cost effective, intelligent, and autonomous with ease-of-use providing better quality of life. Most definitions for smart city involve the use of information and communication technologies (ICTs) to enhance the quality of urban life with reduced cost and resource consumption. Recently, ICT convergence with the Internet of Things (IoT) has been effectively exploited to provide many novel features with minimum human intervention in smart cities. This book describes different components of IoT for smart cities including sensor technologies, communication technologies, big data analytics, and security. The book is organized into five chapters that are described below.

IoT offers smart solutions for cities in terms of governance, economic growth, environmental sustainability, quality of life, transportation, power, and water usage. In Chap. 1, the authors provide an insight into different aspects of smart cities, challenges, and common IoT solutions for future cities.

In Chap. 2, the authors provide an overview of the general classification of communication protocols for IoT networks followed by the analysis of the technical details and specific advantages and limitations of different protocol. Recent protocols for IoT networks are discussed with the comparative analysis of two use cases of IoT and the communication technologies.

Chapter 3, titled Dimension Reduction for Big Data Analytics in Internet of Things, presents an overview of dimension reduction in IoT systems. A discussion on solutions for dimension reduction with focus on principal component analysis is also presented to reduce consumption of energy and computation resources.

The Internet of Vehicles paradigm can play a significant role by providing holistic data exchange between charging infrastructure and electric vehicles (EVs) in emerging smart cities. Large-scale implementation of EVs can impose extra burdens on electric grids making the scheduling essential to optimize the charging process. The authors of Chap. 4 present a profit maximization approach for EV charge scheduling in smart distribution systems by considering the cost and speed of charging at the charging stations.

Recently, blockchain technology is investigated extensively for security and privacy in IoT. Chapter 5 provides an overview of the use of blockchain technology for IoT. First, the authors review the literature for a better understanding of research direction in blockchain for IoT systems. The challenges associated with the deployment of IoT and blockchain for the IoT systems are discussed followed by two case studies on smart homes and food supply chain traceability to show the effectiveness of blockchain technology for IoT.

Kamloops, BC, Canada   Waleed Ejaz
Toronto, ON, Canada   Alagan Anpalagan
September 2018

Acknowledgments

We are very thankful to several people who have worked hard to bring forward this unique resource for helping students, researchers, and practitioners. Our students have contributed in part to the writing of the chapters: D. Vong (Chap. 1), S. S. Sahoo (Chap. 2), M. Basharat (Chaps. 3 and 5), and M. Umer, M. Naeem, and A. Alnoman (Chap. 4).

We would also like to thank Divyaa Veluswamy and Brinda Megasyamalan, Project Coordinators at Springer, who worked with us on the project from the beginning to the successful end. Finally, we would like to thank our respective families for their continuous support and encouragement during the course of this project.

Contents

Acronyms

3GPP	Third generation partnership project
3G	Third generation of cellular networks
4G	Fourth generation of cellular networks
5G	Fifth generation of cellular networks
AMQP	Advanced message queuing protocol
BEV	Battery electric vehicle
BLE	Bluetooth low energy
CALM	Continuous air interference long and medium range
CAM	Cooperative awareness message
CoAP	Constrained application protocol
DDS	Data distribution service
DENM	Decentralized environmental notification message
DSRC	Dedicated short range communication
ETSI	European Telecommunications Standards Institute
EV	Electric vehicle
G2V	Grid-to-vehicle
GHG	Greenhouse gas
GNSS	Global navigation satellite system
GPRS	General packet radio service
GPS	Global positioning system
GSM	Global system for mobile communication
H2V	Home-to-vehicle
HACCP	Hazard analysis and critical control points
HEV	Hybrid electric vehicle
ICE	Internal combustion engine
ICTs	Information and communication technologies
IETF	Internet engineering task force
IoST	Internet of Smart Things
IoT	Internet of Things
IoV	Internet of vehicles
ITS	Intelligent transportation system

ITU	International Telecommunication Union
LAN	Local area network
LPWA	Low power wide area
LR-WPAN	Low-rate wireless personal area network
LTE	Long-Term Evolution
MQTT	Message queuing telemetry transport
PCA	Principal component analysis
PET	Plug-in electric train
PEV	Plug-in electric vehicle
PHEV	Plug-in hybrid electric vehicle
PLC	Power line communication
RFID	Radio frequency identification
RPL	Routing protocol for low-power and lossy networks
SIG	Special interest group
SV	Sensor vehicle
UDP	User datagram protocol
V2G	Vehicle-to-grid
V2H	Vehicle-to-home
V2I	Vehicle-to-infrastructure
V2V	Vehicle-to-vehicle

Chapter 1
Internet of Things for Smart Cities: Overview and Key Challenges

1.1 Introduction

The concept of the `smart city` was introduced as the potential solution to the challenges posed by urbanization. The definitions for smart city involve the use of information and communication technologies (ICTs) to enhance the quality of urban life with reduced cost and resource consumption. Recently, the convergence of ICT in the Internet of Things (IoT) has been envisioned in order to provide novel features with minimum human intervention in smart cities. Many modern cities already have great economy, governance, mobility, and environment. However, implementing IoT into these characteristics will allow them to advance further and improve the outcomes of their operations. Common smart solutions for such operations include but not limited to: traffic management, electricity grids, public transit, businesses, water production and consumption, etc. With the integration of IoT in future cities, large amounts of data from different applications can be generated for various solutions and technologies. It is critical for cities to find the best infrastructure based on the output data in order to deliver reliable, secure, and cost-effective services.

Recently, smart cities have become a big trend as shown in Fig. 1.1. Many of the world's main cities have already adopted the concept of smart cities including Toronto, London, New York, Paris, Seoul, Barcelona, and Shanghai. Most cities that have transitioned are focused on the idea of sustainability. On the other hand, private companies such as IBM, Siemens, and Intel are investing in smart cities as well. Since 2010, IBM's Smarter City Challenge has deployed 700 experts to help 116 cities around the world to address their most critical challenges [1]. Some cities that have participated in the challenge are Ottawa, Rio de Janeiro, Chicago, Kyoto, Helsinki, Copenhagen, and Delhi. Siemens has also deployed its innovative technology on infrastructure projects across the world. In addition, Intel is focusing their technology on the IoT which have the ability to transfer data over a network of devices without the need of human interaction. To summarize, smart cities have become a big trend and many efforts have been made worldwide for it.

© The Author(s), under exclusive licence to Springer Nature Switzerland AG 2019
W. Ejaz and A. Anpalagan, *Internet of Things for Smart Cities*, SpringerBriefs in Electrical and Computer Engineering, https://doi.org/10.1007/978-3-319-95037-2_1

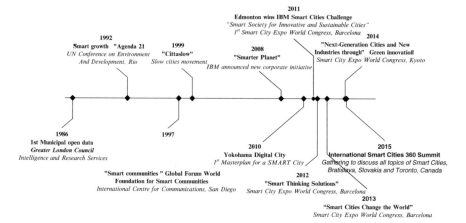

Fig. 1.1 A timeline demonstrating that smart cities are becoming a trend

> Cities basically go out and want to do everything. Free Wi-Fi, meter reading, environmental sensors, information screens, they want them all and they want them for free. Key is to begin by developing a road map of desired services and then prioritize them. (Tormod Larsen, CTO of ExteNet) [2]

The main objective of this chapter is to give readers an insight of different aspects of smart cities and what truly makes them smart. Firstly, this chapter briefly explores the characteristics of smart cities. Secondly, most common IoT solutions for smart cities are summarized. Thirdly, we discuss key challenges ahead for the success of smart cities. Lastly, we wrap up the chapter with a summary and some concluding remarks.

1.2 Characteristics of Smart Cities

In [3], authors developed a model to benchmark and rank smart cities in order to identify strengths and weaknesses in a comparative way. Six key characteristics of smart cities are shown in Fig. 1.2. Further, authors in [4] summarized each characteristic defined by a number of factors. These factors include the following.

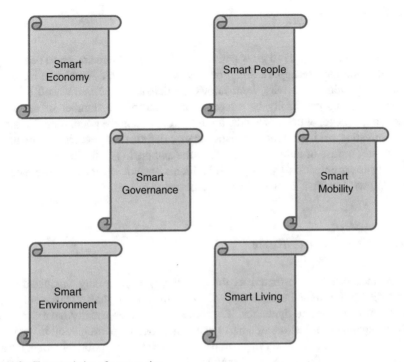

Fig. 1.2 Characteristics of a smart city

1.2.1 Smart Economy

Economy often refers to the wealth and resources of a city, especially in terms of production and consumption of goods and services. The factors that a smart economy should include are entrepreneurship, innovation, trademarks, the flexibility of the labor market, productivity, the integration in the international market, etc. Smart cities are expected to experience economic growth since the population increase will open opportunities to accommodate the needs.

Currently, top 100 urban cities account for 25% of the gross domestic product worldwide [5]. The growing population in cities will result in a greater demand for products and more trade will be required in the international market. Creativity and entrepreneurship are required to accommodate the growth in population. In addition, many smart cities encourage new talent to turn their attention towards smart city projects. Some estimate that $40 trillion will be spent worldwide on new urban infrastructure by 2030 [6], which presents huge innovation opportunities. As a result, there will be a huge availability in the labor market for the upcoming construction of the new urban infrastructures.

1.2.2 Smart People

Smart cities will continue to grow and mature as long as there are smart people and smart technologies to support. Smart people can be measured from some indicators such as education, creativity, innovations, participation, etc. How well educated are the residents can easily be measured in terms of the number of secondary education, college, or university education degrees within the population. Creativity and innovation are important and many smart cities encourage entrepreneurship by a safety net in case of failure. According to the authors in [7], the key success factor of a city is participation by people. If residents are not engaged to co-create and share knowledge, then a smart city goal is bound to fail.

1.2.3 Smart Governance

Governance can be interpreted as the way a city is internally organized. Each smart city is different because they all have their own goals. This results in a new form of economic dynamics. Therefore, smart cities usually have their own form of governance. Factors of smart governance include participation in decision-making, public and social services, transparent governance, political strategies and perspectives, etc. For example, SmartTrack in Toronto is proposed as regional express rail route which is expected to bring economic benefits to the city [8].

1.2.4 Smart Mobility

Transportation is probably one of the most important aspects of a city. Residents need to quickly and efficiently get from one point to another. Most smart cities focus on intracity transportation. For example, New York has a really large underground subway system for people to get around. Seoul has the underground subway system in which passengers can also enjoy the Internet. Barcelona aimed for a more eco-friendly option by using electric vehicles and bicycles. More efficient and greener methods of transportation are inherently considered smarter. In addition, smart cards or access to real-time information for transport systems is a big trend in many smart cities.

1.2.5 Smart Environment

Green communication is one of the common themes in smart cities. Cities want to reduce their carbon footprint. Several efforts have been made in different ways such as upgrading to greener vehicles (electric vehicles) and more efficient waste management. Also, smart buildings can improve the environment and attractivity. For example, in Amsterdam, drinking water is used to cool down buildings by passing it through the building. As a result, buildings are efficiently cooled down without any waste. Essentially, the smart environment includes the following factors: urban infrastructure, carbon footprint, water and energy usage, environmental protection, sustainable resource management, etc.

1.2.6 Smart Living

Residents are the key to a city's development. Improving the quality of life for residents is essential, and attention is required for cultural facilities, health conditions, individual safety, housing quality, education facilities, tourist attraction, and social cohesion. These factors can promote and bring the cultural agenda, tourist guide services for visitors, education, and health.

1.3 IoT-Based Solutions for Smart Cities

There are numerous IoT-based smart solutions worldwide that have been implemented. However, the most common solutions in smart cities are smart grids, smart homes, transportation and traffic management, e-health, waste management, public Wi-Fi, etc. An illustration of the smart city is shown in Fig. 1.3.

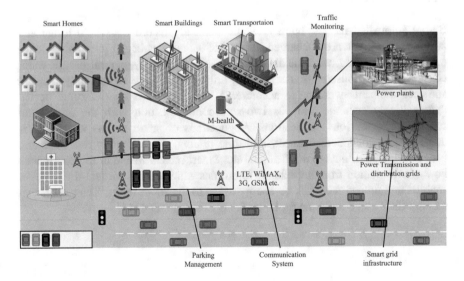

Fig. 1.3 An illustration of a smart city

1.3.1 Smart Grid

The traditional electrical grids use a hierarchical infrastructure in which electricity flows in one direction from a centralized power plant to consumer households. However, the current electrical grids are unable to accommodate the rising demand of the present-day growing population. The infrastructure lacks automated analysis, situational awareness, real-time information, and self-healing. Additional factors that need to be taken into consideration include the capacity limitations of energy generation, one-way communication, and the decrease in fossil fuels [9].

A smart grid is a modernized electrical grid that uses ICTs to manage information between utility and consumers in an automated manner [10]. Furthermore, the grid allows two-way power flow with full visibility and control of assets within the system. The smart grid utilizes advanced sensors deployed throughout the grid to improve fault detection and enables self-healing of the network by re-routing power around the failed equipment and notifying the utility. Robust two-way communication and sensors on the grid will provide customers with access to real-time information of their power usage for better energy management. Generators often have to produce more energy than that required to power cities and prevent blackouts. As a result, a lot of energy ends up being wasted. In a smart grid, sensors monitor power usage and relay that information back to the generators to distribute energy in a more flexible and efficient manner, resulting in larger profits. Overall, the smart grid uses various automated control and IoT to improve reliability, flexibility in the network topology, efficiency, sustainability, real-time pricing methods, etc.

1.3.1.1 Smart Meters

Before automated meters, humans would manually record household usage and customers were often overcharged. In the 1990s, utilities began introducing automated meter reading, with the ability to measure electricity, natural gas, or water consumption of households. As a result, billing costs were significantly reduced and measurements were more accurate. A smart meter is an electronic device that is highly accurate and has two-way communication with the utility and consumers for various applications. Applications of smart meters include anti-theft, remote connect/disconnect, real-time pricing, power-quality measurement, load management, outage notification, etc. Smart meters maintain the continuous flow of information from households to utilities which allows for real-time system analysis and upload feedback on energy usage to the smart meter. Thus, users can monitor their usage real time for better energy management [11].

Table 1.1   Available communication technologies for smart grids [12]

Technology	Spectrum	Data rate	Coverage range	Applications	Limitations
GSM	900–1800 MHz	Up to 14.4 Kpbs	1–10 km	AMI, demand response, HAN	Low data rates
GPRS	900–1800 MHz	Up to 170 Kbps	1–10 km	AMI, demand response, HAN	Low data rates
3G	1.92–1.98 GHz	384 Kbps-2 Mbps	1–10 km	AMI, demand response, HAN	Costly spectrum fees
WiMAX	2.5 GHz, 3.5 GHz, 5.8GHz	Up to 75 Mbps	10–50 km	AMI, demand response	Not widespread
PLC	1–30 MHz	2–3 Mbps	1–3 km	AMI, Fraud Detection	Harsh, noisy channel environment
ZigBee	2.4 GHz, 868–915 MHz	250 Kbps	30–50 m	AMI, HAN	Low data rate, short range

1.3.1.2   Communication Technologies for Smart Grids

There are several types of communication technologies available for smart grid such as power line communication (PLC), ZigBee, WiMAX, third-generation (3G) cellular networks, global system for mobile communication (GSM), general packet radio service (GPRS), etc. However, each type has its own pros and cons. Gungor et al. in [12] briefly summarized them as shown in Table 1.1.

ZigBee is ideal for energy monitoring, home automation (see more in smart homes section), and automatic meter reading. Moreover, deployment of ZigBee devices is low cost and the performance is optimal for demand response, real-time pricing programs, real-time system monitoring, and advanced metering support [13, 14]. However, ZigBee has a short range and cannot cover the larger distance as required by the smart grid.

Wireless mesh is a flexible network consisting of a group of nodes. Data travels through the nodes of the network, where each node acts as a repeater. This enables self-healing characteristics in the network where if a node drops out of the network, then information can be rerouted through other nodes. Smart grid uses this method for self-healing and situational awareness within the grid.

Cellular networks such as GSM, 3G, fourth-generation (4G) cellular networks, and WiMAX are also available for smart grids. Cost can be saved by using already existing communication infrastructure. Further, the data transfer speed of cellular networks is a lot faster compared to other technologies. However, existing cellular networks also share data flow with the customer market which can affect network performance negatively.

PLC utilizes existing power lines to transmit data at high speeds. Since this method uses existing power lines, deployment costs are significantly reduced in many countries. PLCs are the primary choice for electrical grids because they already exist and connected to meters. Also, the security of PLC is stronger than other technologies. However, the transmission medium of the power lines is very noisy and harsh. In addition, the more devices that are connected to the power line in a neighborhood affect the overall quality of the data transmitted as well. In short, PLCs are sensitive to disturbances and are poorly suited for data transmission.

1.3.2 Smart Home

The residential, tertiary, and commercial buildings account for 50% of electricity consumption in Europe [15]. Home and work environments have several energy-consuming units such as lighting, heating, air conditioning, computers, appliances, etc. However, these units are isolated and are often used in a manner with poor energy efficiency and sustainability. As a result, these environments unnecessarily waste a lot of energy. If the units are used in a more coordinated manner, they can provide leverage for energy and cost savings.

The concept of smart homes is a system that constructs an intelligent network which considers each energy-consuming unit as a node. All the units are capable of communicating with each other through the network and can be controlled from anywhere in the household or even remotely through the Internet. The potential of this system is vast as it can be used for security, energy-efficiency, comfort, and convenience.

1.3.2.1 Smart Home Energy Management

Smart homes intelligently manage each device in an energy-efficient manner to reduce the amount of wasted energy [11]. The system will monitor energy usage trends over the time to better manage and automate each device in an optimal manner. For example, the system could operate appliances during off-peak hours such as dishwashers or laundry machines during the night when electricity is cheaper, saving both time and money. Overall, smart homes will provide an energy management system, remote control of devices, timed schedule for usage of appliances, and real-time monitoring among others.

Traditional thermostats operate according to the hysteresis principle. A smart thermostat such as Nest demonstrates to have the ability to learn user behavior patterns. For example, the thermostat will turn off the air conditioning when it detects the users leaving the household and then turn the air conditioning back upon arrival. By auto-scheduling of appliances, the energy consumption can be reduced significantly.

1.3.2.2   Smart Appliances

Smart appliances (lighting, heating, air conditioning, computers, etc.) and systems are often found in newer homes. In contrast, most traditional homes do not have these appliances and systems built in. However, an affordable approach for homeowners is to retrofit them into their households. Smart appliances can be found from many smart home companies such as Notion, Canary, Iris, HomeSeer, Control4, Vera, Savant, Wink, SmartThings, etc.

1.3.2.3   Communication Technologies for Smart Homes

Communication technologies are used to connect devices together in a smart home. ZigBee and Wi-Fi are the most common technologies used among smart home appliances. As mentioned earlier, ZigBee is ideal for energy monitoring and home automation. It has simplicity, robustness, low deployment cost, and easy network implementation.

1.3.3   Transport and Traffic Management

Most highly populated cities experience heavy traffic loads on the road, which ultimately leads to huge amounts of greenhouse gas emissions and waste of money. In 2012, the US Treasury Department reported that approximately 1.9 billion gallons of gasoline have been wasted due to traffic congestion every year at the cost of more than 100 billion in wasted fuel and time [16].

1.3.3.1   Electric Vehicles

Electric vehicles are considered as a potential replacement for the conventional gas-powered vehicles. They are able to reduce carbon dioxide emissions and pollution. Considering that road transport is expected to double by 2050, switching to electric vehicles will significantly reduce the emissions over the long run. Many cities such as Malaga, Paris, Amsterdam, and Barcelona have switched to electric vehicles and installed charging stations across their cities.

1.3.3.2   Intelligent Transport System

An intelligent transport system (ITS) can be defined as a control system that uses ICTs to communicate between vehicles and the highway so as to improve the safety, vehicle, and road efficiency. Vehicles that are equipped with ITS are capable of predicting any vehicle hazards and reduce reaction times to prevent accidents and increase safety. The ITS solutions cover a broad range of situations such as adaptive cruise control, obstacle warning, lane detection, collision notification, etc.

Adaptive cruise control is a system that ensures that a car has a set distance behind another vehicle. Sensors on the front side will monitor the relative speed of the vehicle ahead. The adaptive cruise control system will then adjust the vehicle speed to maintain a safe distance. In the case of a vehicle in front slowing down or another car cuts in front, the system will alert the driver to slow down.

Obstacle warning is another approach that prevents accidents from occurring. This technology uses radar, ultrasound, infrared, and laser to detect obstacles while the vehicle is moving. The driver will be alerted if any obstacle is detected while the vehicle is moving forward or backward.

Lane detection estimates the direction of the road and the position of the moving vehicle along with sensors watching the road. When a vehicle is not properly aligned with a lane, then the driver will be alerted. Furthermore, the vehicle is able to guide the driver back into the lane.

1.3.3.3   Physical Infrastructure

Vehicles are getting safer and smarter. Investing more resources for physical infrastructure of transport management will be more effective in dropping emissions and wasted fuel. For example, traffic is sometimes caused by drivers trying to find the parking spot. However, if sensors are deployed in parking spots, then a system can be implemented to find empty spots. This can help to reduce traffic significantly and drivers will have an easier time finding empty parking spots.

1.3.3.4   Public Wi-Fi

Many smart cities provide free public Wi-Fi which enables any device to connect to the Internet. In addition, residents will be able to access a broad range of citywide services through the network. A large amount of public data will be easily accessible for open data projects. Thus, startups are getting encouragement and ultimately they are improving the city economy. Common applications using data from public Wi-Fi are real-time updates for bus stops, parking availability, monitoring traffic on the highway, etc.

Copenhagen intelligent traffic solutions (CITSs) [17] is a project that installs Wi-Fi access points in a mesh network with the ability to geo-locate Wi-Fi-enabled devices on the street without compromising security. The data is aggregated and anonymized and then fed back to a cloud-based software for city officials to monitor real-time traffic conditions and run simulations. The software can look for patterns and predict traffic behaviors using historical patterns, weather conditions, roadworks, and special events.

1.3.4 Smart Healthcare

Healthcare has become too expensive for many individuals and it suffers from the availability of services, medical errors, and wastes. Every year, there are millions of preventable medical errors that lead to casualties. However, incorporating ICTs within the healthcare sector led to the concept of electronic health (e-health). ICTs help decrease costs and increase efficiency in many healthcare practices. As a result, healthcare facilities are becoming more affordable, and yield better results and increased satisfaction among patients. It encompasses a variety of uses such as communication between patients and doctors, distant diagnostics for patients, electronic medical records, telemedicine, teleconsultants, etc. E-health also removes the need to travel and reduces the costs of medical resources.

The usage of mobile devices such as smartphones and tablets has significantly increased over the past decade. Following the growing trend in mobile devices is the idea of mobile health (m-health) which delivers healthcare services via mobile devices. M-health extends the advantages of e-health to mobile devices. M-health focuses on three important aspects: easy access to services and knowledge, user-oriented, and personalized. A wide variety of services that m-health can provide include but are not limited to health tips and education, health tools, health facility information, medical calculator, clinical and educational use, etc. Numerous mobile health applications have been released for iOS and Android OS devices such as Weight Watchers Mobile, Lose It!, First Aid, Instant Heart Rate, Fooducate, Glucose Buddy—Diabetes Log, etc.

In [18], authors summarized a list of smart solutions in some smart cities worldwide. An expanded list is presented in Table 1.2.

1.4 Challenges Ahead

It can take a decade for cities to transform into smart cities. There are several factors and challenges that are to be taken into consideration before moving forward. The reasons behind why a city may want to transition can gauge how fast it will take. Sometimes, cities want to channel their resources to improve the city for higher quality of life and other times cities may be rebuilding itself after a natural disaster or catastrophe. Following are some major challenges that need to be considered for smart cities.

1.4.1 Planning

Cities are shaped by the inhabitants and understanding the human behavior is critical. For better decision-making, it is essential to investigate urban dynamics,

Table 1.2 Smart solutions in different world cities

Smart city	Smart solutions
Malaga	• Automated meter reading • Electric vehicle • Electric charging stations • Energy efficiency in public and private buildings • Efficient management of public lighting • Smart grid • Development and deployment of renewable energy • Reducing CO_2 emissions • GeCor Program—citizen opinions and complaints
Copenhagen	• Bike lane network • Efficient and reliable integrated public transport • Improvement of water quality • Improved sewage network and water consumption optimization • Development of wind energy • Waste management optimization • Energy efficiency for heating and cooling systems
Paris	• Electric charging stations • Bicycles exchange plan • e-Health: Computer software ELIOS • PROMETHEE for integration of medical records • Issy grid (Smart grid) • Clichy Batignolles • Energy-efficiency refurbishment of housing
Hong Kong	• Citizen-centric online services • RFID in airports • Smart card identification for citizens • Public Wi-Fi • Open Government Data • Big Data Analytics • Paperless Government
Amsterdam	• e-Citizen Participation • Electric vehicle • Electric charging stations • 40% reduction in CO_2 emissions • Increasing energy efficiency of buildings • Smart grid • Introduction of ICT in health • Environmentally efficient waste management • Sustainable district heating
Barcelona	• PROJECT iCity—APP to serve citizens • Electric vehicle • Control and management of urban traffic • Efficient and sustainable urban transport • Centralized heating and cooling network • Environmentally efficient waste management—smart containers • Barcelona Wi-Fi project • Telecare and incident detectors at home

Table 1.2 (continued)

Vienna	• Urban energy systems • Smart grid • Efficient buildings • Reducing CO_2 emissions • Senior Pad • Vienna main station • Citybike Vienna • Energy-saving trains • Citizens' solar power panel
Stockholm	• 50% of domestic hot water from solar energy • Waste management system • Waste water for biogas production • Water management policies • Smart grid • City website for residents (Stockholm.se) • Environmentally efficient waste management • Energy-efficient buildings
Toronto	• Efficient metropolitan urban mobility • Green construction policy • Community broadband network • Cellphone service and Wi-Fi enabled in subway • Smart grid • Bike share Toronto
London	• London online portal • London smart card • Public service networks • Optimization and management of the underground • Metropolitan area with Wi-Fi • Smart grid
New York	• Start-up development for social web • Open government–Open data • Improving public transportation • Application development for services to citizens • Pedestrian spaces for citizens • Public Wi-Fi
Rio de Janeiro	• Control and management of urban traffic • Public transport management • Urban security system—Emergency system control and weather • Integrated health systems
Tokyo	• Solar panels with storage batteries integrated into homes • Bullet trains • Home appliances communicate with each other for efficiency • Smart grid
Vancouver	• Development of electric vehicles • Promoting green transport • Reducing greenhouse gas emissions by 33% • Reduction in water consumption • Environmentally efficient waste management

open data, and residents participation. A common issue with many cities is that they are often in a rush to become a smart city. As a result, projects are often insular, creating an information island that wastes funds because of repeated and redundant construction.

Many cities do not have a master plan or city development plan. It is essential to plan a smart city and act based on a city's needs in order to improve and provide better facilities to residents. Retrofitting existing legacy city infrastructure to make it smarter is another common issue that cities face. There are many challenges when reviewing smart city strategies. One of them is being able to determine the areas that require improvements. Also, integrating isolated legacy systems into the city is very difficult.

1.4.2   Costs and Quality

Choosing between low costs versus high quality has always been a tough decision. Investing in low-cost materials and resources for smart city projects will result in reduced performance and/or quality. On the other hand, higher quality materials and resources often perform better; however, it is only available at a higher cost.

A perfect example of costs versus quality can be deciding which sensors to use. Sensors are one of the primary devices used in smart cities. They are used for smart water and electricity meters, global positioning system (GPS) devices, traffic sensors, parking meters, weather sensors, crowdsourcing, etc. Low-cost ubiquitous sensors can be used in large numbers; however, they produce a low-quality signal and may often require recalibration. In contrast, expensive sensors are more accurate and can be self-calibrated. However, the cost of the expensive sensors will be too high to install for large area coverage.

1.4.3   Security and Privacy

Many smart solutions require the use of ICTs which raises a concern for information security. The technology scale is so large in smart cities that even a small weakness can cause considerable damage. Measures to strengthen this concern include enforcing regulations and laws regarding information safety, implementing information security levels and risk assessment systems, improving the network monitoring capabilities, and strengthening network management. Data that is produced from sensors are used to create effective models. However, the data can be intrusive to some residents making them uncomfortable. As a result, placing sensors on everything may be impractical.

1.4.4 Risks

There are several potential risks that come with smart cities. There are technology risks, operational risks, construction risks, market risks, and policy risks. Each type of risk is summarized in Table 1.3.

Table 1.3 Types of risks involved in smart cities

Type	Description
Technology risk	Risk that new technology does not perform as expected in real-life deployment.
Operational risk	Risk that an operation may not operate to its fullest potential because of the lack of skilled operators.
Construction risk	Risk of unexpected delays or difficulties that can arise during construction.
Market risk	Risk that the market demand for a new service or product is below expectation, leading to a loss-making operation.
Policy risk	Risk that regulatory framework changes, leading to a fall in the profitability of the project.

1.5 Conclusion

Smart cities have become a necessity that addresses the challenges due to rapid urbanization. The solutions highlighted in this chapter demonstrate how cities have tackled these issues to improve the quality of life for its inhabitants. The number of cities worldwide pursuing smart transformation is quickly increasing. However, these efforts face many obstacles in political, economic, and technical aspects. There are several factors and challenges that are taken into consideration before moving forward. Smart city initiatives often require detailed coordination, funding, and continuous support. There must be a return on investments which also presents another challenge. The technical obstacles are another key to ensure security and privacy. In addition, accommodating a proliferation of resources and infrastructures is very important over the long run of all smart city initiatives.

Chapter 2
Communication Technologies and Protocols for Internet of Things

2.1 Introduction

Recently, Internet of Things (IoT) has produced a major paradigm shift in the way devices have communicated with the physical environment. IoT aims at transforming the physical devices into smart objects which can communicate over the Internet. Currently, human-to-human interaction is a dominant paradigm over the Internet; however, IoT proposes a novel emerging paradigm of thought which postulates that any object, identified with a unique identifier will be considered as interconnected. It uses sensors, near-field communications, real-time localization, etc. as a means to interact with the physical environment and serve their purposes. Interconnected devices can help form informed and intelligent decisions in myriad situations such as power grid systems, traffic management, home security and automation, disaster management, etc. Figure 2.1 shows an illustration of IoT network where several devices are connected with different available communication technologies. IoT has also been described as a paradigm that mainly integrates and enables several technologies and communication solutions including but not limited to tracking technologies, wired, wireless sensors, their networks, and exchanged networked communication which in turn lead to a shared next-generation Internet what is also known as Future Internet [11, 19, 20].

IoT network primarily comprises three major constituents: (1) sensors, (2) communication channels, and (3) processing units. The sensors in the devices are responsible for measuring the physical parameters from surrounding environment as and when required. The communication channels serve as the veins of the entire system as they are in charge of transmitting the information collected by the sensors either to the processing unit or to other devices. They filter, compress the information into packages, and send them over suitable channels using specific predetermined protocols. The processing unit is the main brain of the whole system which analyzes the information received and takes decisions.

© The Author(s), under exclusive licence to Springer Nature Switzerland AG 2019
W. Ejaz and A. Anpalagan, *Internet of Things for Smart Cities*, SpringerBriefs in Electrical and Computer Engineering, https://doi.org/10.1007/978-3-319-95037-2_2

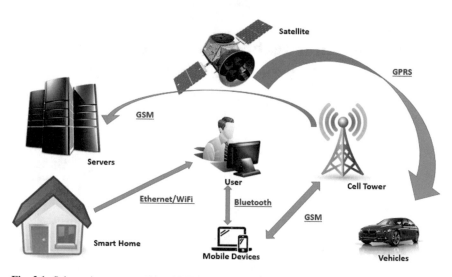

Fig. 2.1 Schematic representation of IoT network with connected devices

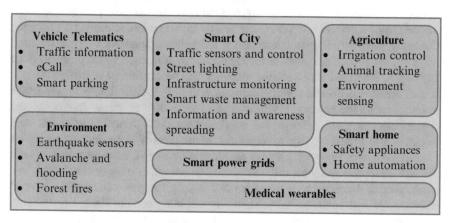

Fig. 2.2 Use cases for IoT

IoT is being used in various scenarios based on the service requirements, data throughput, latency, connectivity, and reliability. The sectors in which IoT is being used are projected to rise exponentially in the years to come. The myriad areas and situations in which IoT devices are being used include vehicular communication, smart grid, medical wearable, smart homes, etc. Figure 2.2 illustrates the different use cases of IoT such as vehicle telematics, agriculture, medical wearables, smart homes, etc.

Communication in IoT can be between devices or from device to base station (processing unit). The communication channels are responsible for compressing the information, packaging them, and ensuring the accuracy of transmitted information and decisions. In this chapter, we will cover various communication methods that have been or are being employed for the transmission and reception of information and decisions to and from devices and base stations. We present an overview of various communication technologies for IoT and technologies adopted by telecom companies such as Nokia, Ericsson, Aeris, etc. to give a holistic idea of the technicalities dealt with while deciding the mode of communication for a particular scenario. We also discussed recent protocols for IoT networks. Along with that, we presented two use-case analyses on intelligent traffic system and disaster management to give practical situations where IoT has been implemented.

2.2 Communication Technologies for IoT Networks

In this section, we will analyze the various communication technologies for IoT networks, their characteristics, advantages, points of differences, and limitations. Communication technologies can be classified into non-cellular and cellular technologies for IoT networks as shown in Fig. 2.3.

2.2.1 Non-Cellular Communication Technologies

The most common non-cellular technologies used for IoT networks include Wi-Fi, Bluetooth, and ZigBee. The others include Z-wave [21], 6LowPan [22], Thread [23], etc. Here, we will present technical characteristics, advantages, applicable use cases, and limitations of each technology.

Fig. 2.3 Classification of communication technologies for IoT networks

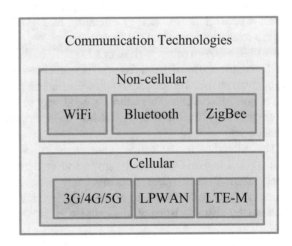

2.2.1.1 Wi-Fi

IEEE802.11 standardized Wi-Fi where the foundational premise of Wi-Fi is inter-operability. An access point from one vendor will nearly always work with a Wi-Fi endpoint from another vendor. Wi-Fi uses full TCP/IP-based protocol, devices communicating via Wi-Fi are known as on the Internet. This means any Wi-Fi-enabled host is by definition part of the local area network (LAN) it joins. This can create security concerns, as IT departments may not be able to secure and manage IoT-connected Wi-Fi endpoints.

2.2.1.2 Bluetooth

Bluetooth is a short-range communication technology and reliable only when endpoints are used within the same room (or within a few dozen meters). Dozens of endpoints in an area are connected back to one master device, which acts as the Bluetooth coordinator/master. Then, that device uses another form of connectivity to communicate to the back end. Then, a large number of endpoints can have inexpensive connectivity using Bluetooth, while only a small number of devices in the system require more expensive connectivity options (like cellular connection). The new Bluetooth low-energy (BLE) or Bluetooth Smart is a suitable option for IoT applications. Since it offers the similar range to Bluetooth, it has been designed to offer significantly reduced power consumption.

2.2.1.3 ZigBee

ZigBee is based on the IEEE802.15.4 protocol, which is an industry-standard wireless networking technology operating at 2.4 GHz targeting applications that require relatively infrequent data exchanges at low data rates over a restricted area and within a 100-m range such as in a home or building. ZigBee offers short-range and low data rate that depends on relaying data between nodes. Data is either trying to get to the endpoint or the access point. Configuring and optimizing mesh networks is a major undertaking. This can be a good way to get wide-area, power efficient coverage. However, low-density, ad hoc mesh networks cannot provide reliable connectivity.

The difference in the various characteristics of each non-cellular technology as discussed above, instinctively, gives rise to their usage in separate scenarios, independent of each other. Similarly, they have their own set of benefits and limitations when subjected to various differences in test conditions. We have summarized characteristics of non-cellular technologies in terms of specifications, operating frequency, range, and data rates in Table 2.1. Further, a comparative study of non-cellular technologies is presented in Table 2.2 with respect to use cases, benefits, and consideration.

Table 2.1 Summary of characteristics of non-cellular communication technologies

Protocol	Specifications	Frequency	Range	Data rates
Wi-Fi	Based on 802.11n	2.4 GHz and 5 GHz Bands	50 m	Gen : 150–200 MBps; Max: 600 MBps
Bluetooth	Bluetooth 4.2 core specification	2.4 GHz (ISM)	50–150 m (Smart/BLE)	1 Mbps (Smart/BLE)
ZigBee	ZigBee 3.0 based on IEEE802.15.4	2.4 GHz	10–100 m	250 kbps

Table 2.2 Comparative study of forms of non-cellular communication technologies

Protocol	Use cases	Benefits	Consideration
Wi-Fi	• Barcode scanners in factories • Connected machines	• Near ubiquitous network coverage in enterprises • Inexpensive chipsets and modules • Can be power efficient, if application and polling rate is designed well	• Friction for third-party devices joining Wi-Fi networks • Provisioning of credentials is difficult
Bluetooth	• Light control • Proximity monitors • Disposable asset trackers (Active RFID)	• Low cost: disposable or competitive product lines • High data rates • Long battery life	• Very short range • Requires key coordination at both endpoint and access point • Needs access point (phone or application-specific device)
ZigBee	• HVAC sensing and control • Lighting control (high density)	• Resilient physical system architecture • Modification or expansion can happen without system disruption • Good power budget if designed correctly	• Short range • Link performance problems • Deployment difficult • Interoperability is often not possible due to configuration differences and key management

2.2.2 Cellular Communication Technologies

The defining features that have made cellular communication protocols competitive enough to be at par with non-cellular modes are:

- **Low battery life:** Even though cellular devices, especially smartphones, are prone to frequent charging, many IoT devices must operate for very long times, often years. They cannot afford to discharge quickly as it may severely hamper the very purpose of their usage. Hence, battery life is an important factor. The sensors used in IoT are low power; however, communication takes up a huge share of power requirement in case of high speed or long-distance IoT communication. The industry target is a minimum of 10 years of battery operation for simple daily connectivity of small packages [24].
- **Low device cost:** IoT connectivity will mostly serve very low average revenue per user with a tenfold reduction compared to mobile broadband subscriptions [24].
- **Low deployment cost:** Deploying low-power wide-area (LPWA) IoT connectivity on top of existing cellular networks can be accomplished by a simple, centrally pushed software upgrade, thus avoiding any new hardware, or site visits.
- **Full coverage:** Enhanced coverage is crucial to many IoT use cases. This has driven the M2M community to look for methods to increase coverage by tolerating lower signal strength than is required for other devices. The target for IoT connectivity link budget is an enhancement of 15–20 dB [25]. The coverage enhancement would typically be equivalent to wall or floor penetration, enabling deeper indoor coverage.
- **Support for massive number of devices:** IoT connectivity is growing significantly faster than normal mobile broadband connections and by 2025 there will be seven billion connected devices over cellular IoT networks [26]. Therefore, LPWA IoT connectivity needs to be able to handle many simultaneous connected devices.

Now, we discuss the available cellular technologies for IoT communication.

2.2.2.1 Third-Generation Network (3G)

Third-generation partnership project (3GPP) third-generation (3G) network's coverage has not grown significantly in the past few years. The capital and effort for the operators deploying this technology have been diverted to their fourth-generation (4G) LTE expansion. 3G networks increase both the download and upload performance for IP data [27]. The equipment and network connectivity for 3G are more expensive than second generation (2G); however, the higher performance better suits evolving applications.

2.2.2.2 Fourth-Generation (4G)-LTE

3GPP fourth generation (4G) is an `All IP` technology. The bits and bytes are transported using IP data packets and control messages are also modified to fully use IP. 4G LTE uses OFDMA technology, hence the devices that use must be multi-mode. LTE networks are designed primarily for broadband communications. They are optimized for high-quality voice and data, including video capabilities. Cellular operators are deploying LTE at different frequency bands because of the allocations of the spectrum they acquired. This use of differing bands is important for companies currently considering LTE deployments. For example, AT&T in the USA is deploying LTE at 1.7 GHz/2.1 GHz and will eventually convert much of their current 850 MHz and 1900 MHz GSM bands to LTE. In the future, Verizon is likely to convert their 800 MHz and 1.9 GHz CDMA deployments to LTE. Sprint intends to deploy LTE at their 800 MHz Nextel frequencies and some unused 1.9 GHz blocks [28].

2.2.2.3 LTE-M

In [27], a new design of IoT system has been proposed that is built from the existing LTE functionalities. It can be deployed using one GSM channel (200 kHz) and can also share spectrum with existing broadband LTE systems. The module cost can be reduced by a magnitude of fourfold with the battery life of up to 10 years using two AA long-life batteries. System coverage is equivalent to 20 dB extension compared to LTE.

2.3 Recent Protocols for IoT

Many standardization bodies and groups are working to provide protocols for IoT including IEEE, ZigBee Alliance, Bluetooth special interest group (SIG), Internet Engineering Task Force (IETF), International Telecommunication Union (ITU), and European Telecommunications Standards Institute (ETSI). In addition, several organizations joined on a single platform called oneM2M to avoid redundancy in the standardization process [29]. The aim of oneM2M is to develop a standard which addresses the need for IoT applications and services. Figure 2.4 shows some recent and diverse protocols for IoT proposed by these bodies and groups.

2.3.1 PHY and MAC Layers

The ultimate success of IoT and M2M is restricted by the standardization. There are many technologies available for IoT and M2M applications including IEEE

Fig. 2.4 Protocols for IoT

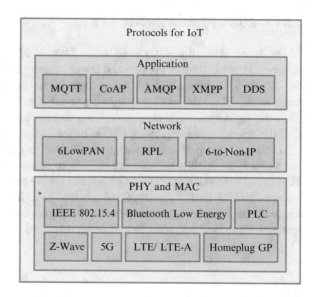

802.15.1, Bluetooth smart or BLE, etc. IEEE 802.15.4 supports the transmission of MAC frames through the physical channel for low-rate wireless personal area networks (LR-WPAN). It provides addressing, data management, and channel access control to allow a single medium to be shared by several IoT devices. However, the probability of colliding frames increases with the number of active IoT devices as the access method used is CSMA/CA. Bluetooth smart or BLE offers short-range communication with low power consumption. The coverage range of BLE is 100 m with less latency and transmission power compared to the classic Bluetooth technology. These features make BLE a good choice for IoT systems. However, these standards need to be connected to the Internet via LTE or Wi-Fi.

The LTE and LTE-advanced (LTE-A) offer higher bandwidth, ubiquitous coverage, mobility support, and plug and play features. In addition, the worldwide deployment of LTE and LTE-A makes it a suitable candidate for IoT and M2M communication. However, they are designed for human-to-human (H2H) communication which makes it both data and power hungry. Thus, a common random access channel (RACH) is used to establish radio bearers before the actual IoT data transmission. The incredible growth of IoT devices results in large signaling overhead as compared to actual data to transmit and can cause congestion on RACH. Therefore, the standard bodies, 3GPP and IEEE 802.16p, have been considering the limitations of cellular systems for IoT and M2M communications. According to 3GPP release 12, LTE now supports Category 0 (or Cat 0) for low-power IoT and M2M devices. Many leading companies are working on LTE Cat 0 and the 3GPP is working on LTE Cat-M in order to support IoT and M2M devices. Despite all these efforts, there are still many challenges need to be addressed for the success of IoT and M2M communication in LTE.

Although there are many add-ons proposed for IoT and M2M communication over LTE and LTE-A, however, major improvements can come with the native support of M2M communication over cellular networks [30]. The 5G cellular networks should satisfy some fundamental requirements for IoT and M2M communications including: (1) support of massive M2M devices, (2) minimum data rate should be ensured in all given conditions, and (3) low latency. Disruptive changes at node and architecture level are required in 5G cellular networks to meet the requirements of new technologies.

The IEEE 1901.2a-2015 is the standard for low-frequency power line communication (PLC) for smart grid applications. Similarly, IEEE 1905.1a-2014 is the standard for a convergent digital home network for heterogeneous network technologies. HomePlug Alliance [31] is a leading industry consortium for PLC and considered as potential communication methodology for IoT applications. For example, HomePlug Green PHY is designed for smart energy/IoT applications with the main focus on low cost, low power, and low data rate. Similarly, Z-Wave is a protocol for home automation networks (HAN) employed by Z-Wave alliance [32]. It has been used widely in smart home applications such as light control, household appliance control, smart energy, etc.

2.3.2 Network Layer

IoT envisions a large number of connected devices or things. It is expected that the huge IPv6 address space, comprising of 2^{128} unique addresses would be an important tool to make this vision a reality. However, traditional IPv6 requires large overhead in terms of code and memory size.

6LoWPAN (IPv6 low-power wireless personal area networks) is the networking technology/adaptation layer that facilitates transmission of IPv6 packets in small link-layer frames such as 127-byte long IEEE 802.15.4 frames, Bluetooth Smart, PLC, etc. 6LoWPAN can connect low-power, IP-driven, and large number of devices which makes it a suitable option for IoT. In addition, 6LoWPAN supports IPv6 neighbor discovery feature that enables the formation of ad hoc networks and mesh under-routing which uses link-layer addresses instead of IP addresses to forward packets. This networking technology facilitates interoperability of IoT devices by providing end-to-end IP connectivity across the ubiquitous Internet by connecting a 6LoWPAN mesh network with IPv6-talking devices using IPv6 edge routers, thus making an excellent use of the already established and widely used Internet technology.

Routing is an important issue in IoT as the devices are low power and battery operated. IPv6 routing protocol for low-power and lossy networks (RPL) developed by IETF offers a versatile routing solution catering to a wide variety of link protocols and link quality parameters. RPL provides a mechanism to support multipoint-to-point traffic (i.e., from IoT devices to a central point), point-to-multipoint traffic (i.e., from a central point to IoT devices), and point-to-point in low-power and lossy networks. With the 6LoWPAN adaptation layer connecting the data link layer to the layers above, ROLL-RPL offers a versatile routing solution catering to a wide variety of link protocols and link quality parameters.

IoT6 is a project on IPv6 and related standards to address the shortcomings and fragmentation of IoT sponsored by the seventh framework programme (FP7) for research of the European Commission [33]. The aim is to exploit IPv6 features for IoT and design a service-oriented architecture to enable interoperability, mobility, cloud computing, and intelligent distribution among heterogeneous IoT devices and applications.

2.3.3 Application Layer

Constrained application protocol (CoAP) is an application layer protocol for IoT applications. It facilitates the integration of IoT with the Internet using the REpresentational State Transfer (REST) architecture (based on request–response model) on top of HTTP. REST provides a lighter way to exchange data over HTTP for IoT applications. CoAP uses user datagram protocol (UDP) over IP in its most typical setting, a fixed binary header, and supports encoding resulting in smaller packets that do not require fragmentation at the link layer. This reduced packet header size and the use of UDP at the transport layer minimizes power consumption and extends the battery life of the IoT devices and helps avoid packet fragmentation at the lower layers.

Message queue telemetry transport (MQTT) is a lightweight messaging protocol for IoT and M2M applications. MQTT uses publish/subscribe mechanism for transmitting data from IoT devices to the server. Unlike CoAP, MQTT is built on top of TCP. MQTT defines three levels for QoS for assurance of data delivery: (1) at most once, (2) at least once, and (3) exactly once. At most once is a best effort service and there is no acknowledgment. On the contrary, at least once acknowledges the receipt of data and retransmits if not received. Exactly once is the case when sender requests to send and starts transmitting data when received clear to send message, and then waits for the acknowledgment. In MQTT, server retains the last message even after sending it to IoT devices and all the new subscribers get this as a first message.

There are other application layer protocols available such as: (1) extensible messaging and presence protocol (XMPP) is used for multi-party chatting, voice, and video calling, (2) advanced message queuing protocol (AMQP) focuses on message-oriented environments and can support reliable communication using TCP, and (3) data distribution service (DDS) is a publish–subscribe protocol that can support real-time IoT applications. Each of these protocols has its own objective and may perform well in a particular scenario. Thus, application layer protocols can be reconfigured according to the applications and user requirements.

2.4   Study of Communication Technologies Through Use-Case Analysis

We have selected two use cases to study the practical implications of using IoT devices. The use cases have been selected carefully because of the difference in their nature, occurrence, dynamic qualities, and the area of coverage. First use case is on `Intelligent Traffic System` (ITS) which requires a continuous transmission of sensor data round the clock, covers a small area, needs short distance, however, quick response from control station, and is more real time. Second use case is on `IoT for Disaster Management` where the area covered is very wide, communication needs to be long distance, precise, and over a relatively short period of time. It needs to be triggered as and when required in time of distress and deals with a more sensitive issue which requires more human intervention than the first use case. Now, we analyze how the communication technologies and protocols differ in both the cases.

2.4.1   Use Case 1: Intelligent Traffic System (ITS)

The ITS is implemented using roadside units (RSU) with friction monitoring, vehicles with environmental sensors, and a database for data transfer through different platforms. The system is able to collect sensor data from stationary RSU stations or moving vehicles and store it in the database. The RSU can take images of the road section with a stereo camera and calculates the road weather type and from the lookup table, the system estimates the road friction based on the measurements. The RSU is able to send measurements to vehicles through V2X communication using cooperative awareness message (CAM) and decentralized environmental notification message (DENM) and to the database using 3G mobile connections. Vehicles can communicate through V2X communication using CAM/DEMN with RSU and other vehicles nearby. In addition, vehicles communicate with the database using mobile 3G connections.

The database is used to store all the measurements from the vehicles and RSU. The intelligent traffic system database contains weather information from environmental sensors combined with data from vehicle sensors. The ITS is implemented by combining RSU and vehicle sensors together with two communication channels vehicle-to-vehicle (V2V)/vehicle-to-infrastructure (V2I) communication is implemented using IEEE 802.11p communication. Vehicles or RSU can send CAM/DENM messages to other nearby units/vehicles. One message can contain the global navigation satellite system (GNSS) position of the sender, the message type, and an actual message. In the ITS, the message contains weather warnings together with the position measurement and the friction measurement. The receiver of the message can calculate its distance to the measurement point and provide a warning to the driver if any action is needed.

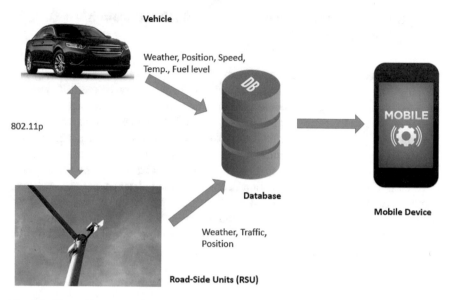

Fig. 2.5 Illustration of intelligent traffic system

In addition to the I2V message, RSU and vehicles communicate through the database. Some other modes of communication among the vehicles include [34]:

- Dedicated short-range communications (DSRC) provides communications between the vehicle and the roadside in specific locations (e.g., toll plazas).
- Wireless communication systems dedicated to ITS, vehicles, and traffic telematics will provide network connectivity to vehicles and interconnect them. Using radio bands requires adequate harmonized standards which are under development for the bands 5 GHz and 63 GHz.
- Continuous air interface long and medium range (CALM) provides continuous communications between a vehicle and the RSU using a variety of communication media, including cellular, 5 GHz, 63 GHz, and infrared links. CALM will provide a range of applications, including vehicle safety and information as well as entertainment for driver and passengers (Fig. 2.5).

2.4.2 Use Case 2: Disaster Management

In disaster management, it is apparent that a number of teams and individuals from multiple, geographically distributed organizations (such as medical teams, civil protection, police, fire and rescue services, health and ambulance services, etc.) will

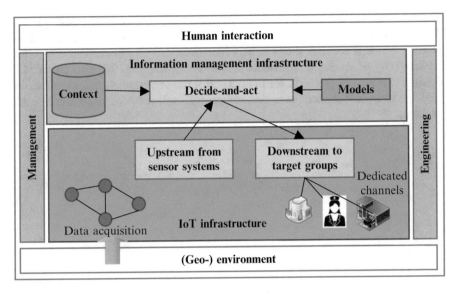

Fig. 2.6 Illustration of IoT platform for disaster management [36]

be required to communicate, cooperate, and collaborate—in real time—in order to take appropriate decisions and actions.

The challenges during and after a disaster are many. One of them is that many small communities do not have the resources, manpower, and expertise to develop a set of requirements to assist them in managing their activities as they pertain to the emergency. Along with that, there are huge loopholes in resource management and communication among the rescue personnel. The combination of social networks in an IoT environment and its analysis through the use of social network analysis stands as an interesting approach for dealing with disaster and crisis management scenarios [35] (Fig. 2.6).

2.5 Conclusion

In this chapter, we provided an overview of two broad types of communication technologies for IoT networks, i.e., non-cellular communication and cellular communication. Non-cellular communication technologies are profitable and advisable when the coverage area is less and economical solutions are required. However, they tend to fail when a larger geographical area has to be covered. Since most of the devices continue to incorporate cellular features, they are the next big feature in IoT paradigm. Among cellular modes, 3G has been dominating the scenario; however, now LTE and LTE-M have come up with their own set of advantages to makeup for the shortcomings of 3G. The large range of frequencies that the LTE deals with implies that the LTE radios must be able to jump from one band to another quickly, which if not shall prove to be a hindrance. LTE is an excellent choice for the longevity of service with module prices slowly dropping. We also provided a detailed overview of protocols for IoT networks. Finally, we analyzed two use cases from the viewpoint of communication technologies and protocols to be used in each of them.

Chapter 3
Dimension Reduction for Big Data Analytics in Internet of Things

3.1 Introduction

Internet of Things (IoT) is providing and expected to provide solutions to real-life problems with an objective to improve quality of life. The key of the IoT is connectivity of physical objects through the Internet through a wired or wireless technology. IoT provides a wide range of solutions such as environment monitoring, intelligent transportation system, e-health, smart grid, smart homes, etc. [19]. The design of sophisticated IoT solutions and its adaptation is leading to a large number of IoT devices. According to a forecast by Ericson experts, there will be 29 billion connected devices by 2022 [37]. These large numbers of devices generate a large amount of data over the Internet in various different formats such as plain messages, images, audio, and video. The information obtained from the data collected by billions of IoT devices can have a significant impact on future living in smart cities and businesses. The heterogeneous data generated by multiple sources can provide detailed insights about physical objects, users, and environment.

Some of the key features of IoT include low-power devices, distributed nature, limited computational power, and heterogeneous data. Considering the low-power IoT devices, it is challenging to read, write, and process a large volume of data. Thus, we need efficient and robust methods to deal with a large volume of data with given constraints on IoT devices. Also, there is a need to integrate heterogeneous data generated by different types of IoT devices. Lastly, the key information needs to be extracted from a large amount of data. The analysis of data generated by massive IoT devices can be done by several different tasks such as dimension reduction, classification, regression, clustering, outlier detection, etc. [38]. Considering the features of IoT, we need new algorithms and methods for data analysis in order to reduce the communication and computation cost for data transmission and task-specific design change.

In this chapter, we will focus on dimension reduction which consists of various machine learning techniques to reduce the number of variables. This consists

© The Author(s), under exclusive licence to Springer Nature Switzerland AG 2019
W. Ejaz and A. Anpalagan, *Internet of Things for Smart Cities*, SpringerBriefs in
Electrical and Computer Engineering, https://doi.org/10.1007/978-3-319-95037-2_3

Fig. 3.1 Categories of
dimension reduction

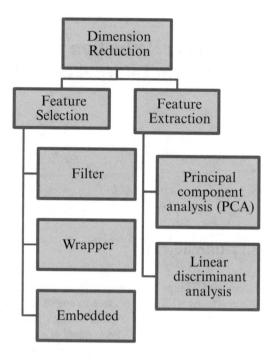

of feature selection and feature extraction as shown in Fig. 3.1 [39]. In feature selection dimension reduction techniques, the objective is to find a subset from a multi-dimensional dataset. The key techniques in this category include filter, wrapper, and embedded. On the other hand, in feature extraction, few dimension data is extracted from the multi-dimensional data. The techniques included in this category are principal component analysis (PCA) and linear discriminant analysis. Mainly, we provide an overview and categories of dimension reduction. We present several challenges associated with dimension reduction in IoT systems. Also, we discuss several examples of dimension reductions using different techniques and approaches. Lastly, we highlight open research issues on dimension reduction in IoT systems.

Rest of the chapter is organized as follows: Sect. 3.2 provides challenges associated with the dimension reduction in IoT systems. Section 3.3 presents several solutions for dimension reduction in IoT systems with examples. Lastly, conclusions are drawn in Sect. 3.4 and future research issues are highlighted.

3.2 Related Work

In [40], authors proposed a two-layer dimension reduction solution for intrusion detection in IoT backbone networks. These layers consist of component analysis and linear discriminate analysis of dimension reduction. Further, authors investigated two-tier classification module: utilizing Naïve Bayes and certainty factor version of K-nearest neighbor to identify suspicious behaviors. Authors in [41] proposed a new approach based on PCA for big data analysis. The proposed scheme can provide an exact solution when the linear regression approach is used in data analysis afterward. In [42], authors proposed a crowdsourcing-based framework for better understanding of the data generated by social users based on IoT. Dimension reduction reduces the amount of data to be stored as well as the communication cost. A framework for big data reduction in IoT is proposed in [43] at the customer end. Authors also presented a business model for end-to-end data reduction in enterprise applications. In [44], authors proposed a full-view area coverage in camera sensor networks. It is shown that minimum number full-view area coverage can be reduced to minimum number full-view point coverage by selecting a particular full-view ensuring set of points. Authors proposed greedy algorithm and a set-covered-based algorithm based on the study of geometric relationship between the full-view coverage and traditional coverage. Authors in [45] presented a solution for real-time data reduction at the network edge. The proposed solution automates the switching between different data handling algorithms. Three variants of proposed algorithm are presented based on perceptually important points concept. An ϵ-kernel dataset concept is proposed in [46] that represent a large information from wireless sensor networks by a small subset of data. The information loss rate of proposed algorithm is less than ϵ which is an arbitrarily small value. Authors proposed distributed algorithms (accurate algorithm and the sampling-based approximate algorithm) to minimize ϵ-kernel dataset to save energy and computation resources.

A summary of the related work on dimension reduction in IoT systems is given in Table 3.1.

3.3   Solutions for Dimension Reduction in IoT

Data analysis becomes hard when there are too many variables involved. There can be different scenarios which we can come across while doing data analysis such as:

- We can explore correlation in the variables involved in Big data analysis for IoT.
- We may decide to analyze complete data which needs more computational power and complexity.
- We need to come up with the methods to find most important variables in data.

As mentioned earlier, dimension reduction techniques are key to get rid of data with many variables. Dimension reduction refers to the process of converting a set of data having vast dimensions into data with lesser dimensions ensuring that it conveys similar information concisely. This can be done by using machine learning techniques to obtain better features for classification or regression task. For example, for a dataset of n dimensions, where n is very large in the order of 100s. We can reduce this dataset to k dimensions, where $k < n$. The reduced number of dimensions can be extracted directly through a filtering process. On the contrary, the dimensions can be reduced by the combination of multiple dimensions such as

Table 3.1 Summary of related work on dimension reduction in IoT systems

Ref.	Year	Objective	Requirements	Solution	Applications
[40]	2016	To map the high-dimensional dataset to a lower one with lesser features to address limitations due to dimensionality	Malicious activities within IoT networks is critical for resilience of the network infrastructure	Component analysis and linear discriminate analysis of dimension reduction module	Security in IoT backbone network
[41]	2016	Reduce dimension reduction for big data analytics	–	Modified PCA	–
[42]	2017	To make full use of the collective wisdom of social users based on IoT	Reduces the amount of data to be stored as well as the communication cost	A crowdsourcing-based framework for better understanding of the data generated by social users based on IoT	–
[43]	2016	To propose a framework for early data reduction at customer end and present a business model for end-to-end data reduction in enterprise applications	Enabling secure data sharing lowering the service utilization cost preserving privacy of customers	Authors proposed a framework for early data reduction at customer end	Enterprises
[44]	2016	Minimize the number of cameras	Guarantee the full-view coverage of a given region	Proposed greedy and set-cover-based algorithms	Camera sensor networks
[45]	2015	Data reduction of time series by applying different data reduction methods with adding notable delay	Real time	Three variants of proposed algorithm are presented based on perceptually important points concept	Edge computing
[46]	2017	Minimize ϵ-kernel dataset	Represent a large information by a small subset of data	Proposed distributed algorithms including accurate algorithm and the sampling-based approximate algorithm	General for all IoT applications

weighted averages or new dimensions can be formed. There are several advantages of using dimension reduction in big data analytics for IoT systems such as:

- Dimension reduction can be helpful to reduce the computational power for data analysis as well as required storage space.

Fig. 3.2 Redundant dataset
for relation of cm to inches
[47, 48]

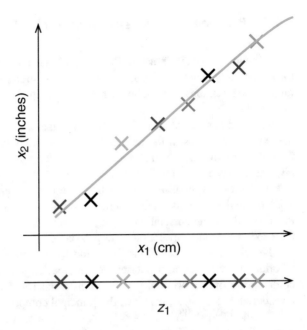

- The communication cost can also be reduced by using dimension reduction in Big data analytics for IoT systems. Data with fewer dimensions will cost less communication overhead and existing algorithms can be applied to the fewer dimension data.
- Data visualization is much more easy with the fewer dimension data. The data patterns can be observed more clearly.
- Noise reduction.

For the illustration purposes, we provide an example of redundant dataset for different units of the same attribute, i.e., centimeters and inches for the measurements [47, 48]. This data is highly correlated and we can combine correlation into a single attribute (Z). Figure 3.2 shows a 2D feature vector (X and Y dimensions). Both dimensions are providing similar information and can be source of noise when machine learning algorithms will be applied. Therefore, it is important to convert 2D data as a 1D vector (Z dimension). There are several existing methods to reduce dimensions such as [49, 50]:

- Random Forest
- Backward Feature Elimination
- Factor Analysis
- Principal Component Analysis (PCA)

3.3.1 Principal Component Analysis (PCA)

In this chapter, we will focus on PCA as a dimension reduction technique. PCA is the most trendy technique for the linear transformation [51–53]. PCA is used in vast range of applications such as smart homes, intelligent transportation system, stock market, etc.

In PCA, a new set of variables called principal components can be derived from the existing variables in the dataset [54]. These principal components are derived from the best variation of the main dataset. The projections along the directions of high variations are retained whereas the directions of low variations are discarded. The more variety confined to the first component results in more information available in the component. One of the major objectives of PCA is to detect patterns in data by detecting correlation among variables. The dimensions can be reduced if there exists a strong correlation among variables.

The principal component can be found by a linear combination of normalized of original predictors in a dataset. A principal component is a normalized linear combination of the original predictors in a dataset. Here, a set of predictors is considered as S^1, S^2,....,S^p. Now, the principal component can be written as $C^1 = \Phi^{11}S^1 + \Phi^{21}S^2 + \Phi^{31}S^3 + ... + \Phi^{p1}S^p$, where the first component is $C1$, Φ^{p1} is the weight also called loading vector for the first component $C1$. The sum of all weights is equal to 1. These weights are defined to set the direction of the first component along with the more variations in the original dataset. S^1, S^2, ..., S^p are predictors with zero mean and standard deviation one.

Steps of PCA

- Data standardization: It is important to standardize data prior to using PCA since the principal components are susceptible to the scale of measurements.
- Then, we can find correlation matrix or covariance matrix from which eigenvectors and eigenvalues can be extracted.
- We then need to arrange eigenvalues in descending order. Further, we select k eigenvectors that represent the largest eigenvalues.
- From the eigenvectors, we can construct the projection matrix M.
- Finally, the original data can be transformed to find k-dimension (reduced dimensions) via projection matrix M.

Case Study: Iris Data We consider a public iris dataset [55], where the flowers belong to three different species: setosa, versicolor, and virginica. For each flower, we have four measurements (in cm): sepal length, sepal width, petal length, and petal width. The data points are in four dimensions. We applied PCA to this data to identify the combination of attributes (principal components, or directions in the feature space) that account for the most variance in the data. Figure 3.3 shows the plot of different samples on the first two principal components, which show that they contain most of the information. After calculations, it is concluded that these two principal components contain 95.8009753615 percent of the original information.

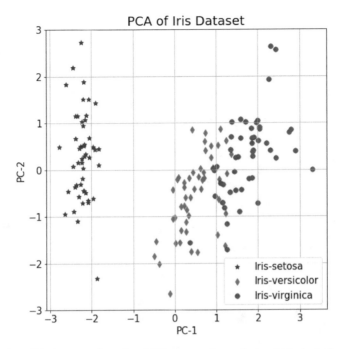

Fig. 3.3 PCA of iris dataset to show plot of different samples on the two first principal components

3.4   Conclusion and Future Work

The sophisticated and ubiquitous IoT applications drive the growth of a large amount of data. It is hard to analyze the data generated by heterogeneous devices and data can consist of multiple dimensions. Dimension reduction is one of the key methods to reduce the number of variables in a given dataset. We presented related work on dimension reduction in IoT systems. We then provided a detailed discussion of the available solutions for data reduction in IoT systems. The focus is given to dimension reduction. We presented a case study on IRIS data for better understanding of the PCA. There are still many open challenges which need to be investigated. It is important to investigate that how communication cost can be influenced by the dimension reduction. Also, how communication cost will effect if data is partitioned vertically. For IoT systems, the number of features can vary as data comes from heterogeneous devices. New algorithms need to be designed to deal with the dynamic nature of IoT systems.

Chapter 4
Internet of Things Enabled Electric Vehicles in Smart Cities

4.1 Introduction

The ubiquity of electric vehicles (EVs) in future smart cities demands efficient and intelligent charge scheduling techniques. Smart grids, on the other hand, are equipped with sensors and meters that significantly assist in the supervision and control of the charge scheduling process. Thus, integrating both EVs and smart grids in a larger IoT entity can facilitate the vast deployment of EVs towards a green and intelligent transportation era.

Conventional motor vehicles mostly depend on the energy of fossil fuels. At the present time, fossil fuels are considered as the worlds' primary energy resources. All technological advancements in the fields of agriculture, transportation, industry, etc., depend heavily on fossil fuels. Global transportation is one of the major sources of fossil fuel consumption such as gasoline, petrol, diesel, etc. Moreover, fossil fuels are considered as non-renewable energy resources which may eventually run out in the near future. Therefore, it is necessary to take serious steps to reserve fossil fuel resources for a stable energy future. Due to the aforementioned reasons, people are now moving towards the diversification of energy resources, particularly for daily transportation needs. According to the US Climate Action Report in 2010 [56], approximately 28% of the total greenhouse gas (GHG) emissions in the USA comes from the transportation sector that includes cars, buses, motorcycles, aircrafts, ships, trains, etc.

Burning fossil fuels is the largest source of GHG due to the emission of carbon dioxide, CO_2. Moreover, it negatively affects global warming that is impacting the entire ecosystem on the planet. For instance, there has been a phenomenal increase in the observed temperature during the second half of the twentieth century, solely because of the GHG emissions. Furthermore, carbon monoxide, nitrogen oxide, and hydrocarbons are released when fuel is burned by the internal combustion engine (ICE) inside vehicles, and emitted into the atmosphere through the vehicle's tailpipe. In addition to the detrimental impact on health, motor vehicle pollution also

© The Author(s), under exclusive licence to Springer Nature Switzerland AG 2019 39
W. Ejaz and A. Anpalagan, *Internet of Things for Smart Cities*, SpringerBriefs in
Electrical and Computer Engineering, https://doi.org/10.1007/978-3-319-95037-2_4

GHG Emissions

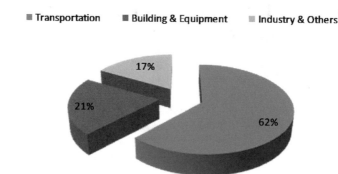

Fig. 4.1 Greenhouse gas emissions in Seattle, USA

contributes to the formation of acid rains. Figure 4.1 depicts the GHG emission in Seattle, USA, and shows that the transportation sector accounts for 62% of GHG emissions.

One way to significantly reduce the pollution incurred by motor vehicles is the usage of green EVs instead of conventional ICE vehicles. On the other hand, electricity is a clean form of energy that can be easily and effectively transformed from one energy form to another. Nowadays, most researchers focus their intensions and objectives towards the vehicles driven by electric motors instead of ICEs. An EV is basically driven by electric energy that is stored in a rechargeable battery (or series of batteries). Currently, EVs are gaining considerable popularity in many developed countries. However, EVs are still used in small scale due to the challenges related to charging efficiency and cost. These challenges can be considerably resolved by the careful scheduling of EV charging, and placing charging stations along the service areas. The advantages of using EVs can be summarized as follows:

- **Clean Environment:** EVs are environmentally friendly and operate without emitting GHGs that impact the health of humans and other living organisms.
- **Economic Performance:** EVs, which use electricity for propulsion instead of fossil fuels, are considered as more economical since the price of electricity is less than that of fossil fuels.
- **Quiet Vehicles:** EVs produce less noise and help reduce the noise pollution.

There are many pros and cons to be considered about EVs. For instance, EVs have faster acceleration but less capability for long distances. Moreover, they produce no exhaust but, however, require long charging times. Moreover, EVs can offer more time flexibility in charging and discharging by introducing the concept of vehicle-to-Grid (V2G). V2G is defined as the capability of returning stored electric energy to the grid from the vehicle's battery. In other words, an EV acts as both a controllable load and distributed storage device. By connecting unused EVs to the electric grid, the batteries of these EVs can provide energy during peak load times, and thus increase the reliability of the grid [57].

4.2 EV Charge Scheduling and Charging Techniques

Charge scheduling (charging and discharging strategies) is essential to avoid grid congestion. Efficient scheduling ensures proper operation of the distribution system. The goal of scheduling is to allocate energy from available resources to where the energy is needed while maintaining the optimal operation of the system without overloading or congesting the main grid [58]. Therefore, scheduling in smart distribution systems helps minimize the operational costs by reducing the electricity bills. Furthermore, smart scheduling plays a significant role in establishing the intelligent transportation system (ITS), where both communication and computing meet in vehicles and charge stations to maintain efficient and reliable performance regarding energy and cost [59]. The central cloud in the Internet of Vehicles (IoV) paradigm will have a holistic view about energy availability and charging demands to optimize the charging process in stations and vehicles.

Different types of EVs are available such as the Plug-in EV (PEV), Plug-in Hybrid EV (PHEV), Hybrid EV (HEV), Sensor Vehicle (SV), Battery EV (BEV), and Plug-in Electric Train (PET). PHVs and PHEVs are EVs with rechargeable batteries that provide power to operate the vehicle. These batteries can be fully recharged by connecting to an external power supply. However, PHEVs have both ICE and electric motors for propulsion.

The research area of EVs has been extensively studied recently. In [60], a scheduling scheme is presented to decrease the peak electricity demand and reduce the electricity bill using SVs and PHEVs. In [61–65], the authors developed techniques for minimizing the peak hour electricity demand by allowing communication between PEVs and the electric grid. The authors in [66, 67] considered that existing power infrastructure will be affected by the increase in the number of the EVs. In [68], the authors introduced a method for achieving maximum fuel economy and minimizing the environmental pollution using HEVs.

Charging and discharging techniques can be classified as vehicle-to-grid (V2G), grid-to-vehicle (G2V), vehicle-to-vehicle (V2V), vehicle-to-home (V2H), and home-to-vehicle (H2V). V2G describes a system in which plug-in EVs such as BEVs and PHEVs communicate with the electric grid to exchange services such as buying or selling electricity from/to the grid with variable charging rates. Research in this field of study involves several categories of objective functions regarding charging and discharging. The major categories include maximizing profit and generating revenue, minimizing cost and power loss, charging station placement, scheduling of EV charging, minimizing pollution, peak clipping and valley filling, and V2G power flow. In [57, 58, 69–74], the authors proposed V2G for discharging EVs through the main grid. Using this technique, customers can generate revenue by selling power to the grid, and help provide power during peak hours. In [75], a system architecture is designed for efficient control of load balancing in EVs at charging stations using V2G technology. In [76], a method for the optimal placement of charging stations in smart cities using V2V is proposed. Furthermore, combining wind power generation with V2G technology helps reduce the intermittency of wind power and lead to more sustainable development [74]. The EVs in a parking garage can be utilized as a dynamic energy storage facility to compensate for the variability of renewable energy resources [77].

Integrating the massive number of vehicles into the power grid might incur other issues and challenges. A stochastic model based on queuing theory for PEV and PHEV charging demands is studied in [78]. This study highlighted the impact of the large-scale integration of EVs on the power grid. An integrated rapid charging strategy that considers both traffic conditions and status of the power grid is introduced in [79]. However, rapid charging could degrade the power system performance especially during peak hours.

Minimizing the operational cost of EV charging has been considered by several researchers using different approaches such as scheduling [57], and shifting the peak load to valley areas where the real-time pricing is low [70]. Moreover, selling electricity to the power grid using V2G can bring profit to vehicle owners. However, this requires efficient energy management in regard with available renewable energy resources [80], operational costs and energy losses [72], and parking lot allocation [81]. In [70, 82], load management approaches are presented using peak clipping (i.e., avoiding EV charging during peak hours) and valley filling (i.e., incentives on charging during off-peak hours). In this manner, customers can generate revenue by selling power to the grid through discharging their EVs, and using renewable energy during peak hours. Similarly, during off-peak hours customers get incentives for charging EVs within a prescribed time limit.

Scheduling algorithms have been proposed to minimize costs and peak electricity demands by considering the factors of fuel pricing, electricity demand, and vehicle characteristics [60]. In addition, the accurate placement of charging stations plays a significant role in improving the performance of EV charging by serving more EVs in less amount of time [83–85].

4.3 Renewable Energy for EV Charging

The most common renewable energy resources are the wind and solar energy. However, other energy resources such as the motion of water, the carbohydrates in plants, and the warmth of the earth can also be exploited to satisfy the energy demands in a sustainable manner. Future EVs need to be fast in both speed and charging time. Current technologies allow EVs to be recharged within a couple of hours. Although this time seems to be short for an EV, it is still very long compared to traditional vehicles. A fast charging mechanism can be envisioned as follows: when a car comes to recharge, a robotic system removes the discharged battery from the car and places it on a conveyer belt. This belt takes the battery to an underground battery recharging unit and replaces it with a fully charged battery. The whole process is done in a couple of minutes which seems very promising.

Scheduling can be achieved either in a centralized or decentralized manner. The centralized model for EV scheduling is effective in reducing the total cost and peak-to-average ratio of load. A decentralized system is the one in which the entry of peers is not regulated, i.e., any peer can enter or leave the system at any time. However, in a decentralized system, there is no centralized authority that makes decisions on behalf of all parties. Instead, each party, also called a peer, makes local and autonomous decisions towards its individual benefits which may possibly conflict with other peers. Moreover, peers can directly interact with each other and share information or provide services.

4.4 Smart Distribution Systems

In real life, a number of constraints are involved in the charging of EVs. Some of these constraints are listed below:

- Maximum and minimum output power limits
- Grid capacity
- Charging limit in batteries
- State-of-charge (SOC) of the battery and the charging rate
- Intermittent supply of renewable energy resources

To cope with the aforementioned constraints, developing smart scheduling schemes is essential to dynamically allocate power for EVs taking into account the charging duration, battery limit, grid conditions, and costs. Moreover, integrating these constraints in the IoT paradigm will make the entire charging process more efficient and reliable. In this section, the problem of EV scheduling in smart distribution systems is presented.

4.4.1 Smart EV Scheduling: A Case Study

The goal of using smart distribution systems is to optimize the scheduling of EV charging. There is different cost associated with different charging levels. For instance, the cost of slow charging is very low but, however, takes more time than other charging levels. Whereas the cost of very fast charging is higher than that of other charging levels but, however, requires much less time. The difference in charging times is due to the amount of charging power, i.e., the more the charging power, the less the time required to charge the vehicle. The system model for EV scheduling is shown in Fig. 4.2.

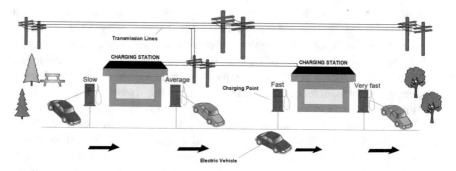

Fig. 4.2 Electric vehicles scheduling system model

Different charging levels depending on the desired charging time are listed below:

- Level 1 (L-1) charging is the slow charging and the vehicle battery is charged by applying 120VAC/16A for a 1.92-kW charging power using the on-board battery charger. The charging time required for full capacity is about 10 h [76]. EV charging that takes place at homes can be considered as level 1 charging, where the charging process is completed during the night.
- Level 2 (L-2) charging is called the standard charging. The vehicle battery is charged by applying 208V-240VAC, 12A-80A for a 2.5–19.2-kW charging power using the on-board battery charger. In L-2 charging, the time required to fully charge an empty battery is 6–8 h. The best implementation of L-2 charging is at places where the user stays for a long time, e.g., at work.
- Level 3 (L-3) is DC fast charging, where charging is done by applying up to 200A for a 75-kW charging power using off-board chargers. The time required for fast charging is about 30 min, whereas the charging price is high [76].
- Level 4 (L-4) is DC very fast charging, where charging is done using up to 400A for a 240-kW charging power using off-board chargers. The very fast charging is required at places where the user cannot wait for a long time, e.g., at public charging stations. The time required for charging is about 15 min with very high pricing [76].

The time horizon T is divided into discrete time slots (15-min time slot), i.e., the time required for charging an EV with very fast charging is 15 min (one time slot) and the time required for the fast charging is 30 min (two time slots). Similarly, the time required to charge an EV using average charging is 45 min (three time slots) and the time required for slow charging is 60 min (four time slots). We always ensure that the number of time slots is greater than or equal to the number of EVs, otherwise the solution will not be feasible. The table provides an overview of the notations used in the problem formulation.

The EV scheduling problem which is a binary integer linear programming problem aims at maximizing the total profit through the scheduling of slow, average,

Notation overview

Symbol	Description
T	Total time
N_v	Number of electric vehicles
c^s	Cost of slow charging
c^a	Cost of average charging
c^f	Cost of fast charging
c^{vf}	Cost of very fast charging
t^s	Time for slow charging
t^a	Time for average charging
t^f	Time for fast charging
t^{vf}	Time for very fast charging
w_i, x_i, y_i, z_i	Decision variable $w_i, x_i, y_i, z_i \in \{0,1\}$, $\forall i \in \{1,2,3,.....N_v\}$
$L-1$	Level 1 Charging power 1.92 kW
$L-2$	Level 2 Charging power 2.5–19.2 kW
$L-3$	Level 3 Charging power 75 kW
$L-4$	Level 4 Charging power 240 kW

fast, and very fast charging levels. This problem is quite similar to a knapsack problem and is an NP-hard (Non-deterministic polynomial time hard). The binary integer variables $w_i, x_i, y_i, z_i \in \{0,1\}$, $\forall\ i \in \{1, 2, 3, ..., N_v\}$ are considered as the decision variables for the slow, average, fast, and very fast charging levels, respectively. For instance, if w_i is "1," then the EV is under slow charging, and if z_i is "1," then the vehicle is charged using the very fast charging level, etc. The optimization problem is formulated as:

$$\min_{w_i, x_i, y_i, z_i} : \sum_{i=1}^{N_v} \left[w_i c^s + x_i c^a + y_i c^f + z_i c^{vf} \right]$$

Subject to:

$$C_1: \quad w_i + x_i + y_i + z_i = 1, \qquad \forall i \in \{1, 2, 3, ..., N_v\} \quad (4.1)$$

$$C_2: \quad \sum_{i=1}^{N_v} w_i t^s + x_i t^a + y_i t^f + z_i t^{vf} \leq T$$

$$C_3: \quad w_i, x_i, y_i, z_i \in \{0, 1\},$$

where c^s, c^a, c^f, and c^{vf} represent the cost of slow, average, fast, and very fast charging, respectively. N_v denotes the total number of EVs available for charging. t^s, t^a, t^f, and t^{vf} denote the time required for slow, average, fast, and very fast charging, respectively. It means that the service provider must ensure that all the EVs and batteries are charged before that time limit. The objective is to maximize the total profit as shown in (4.1) by optimally minimizing the cost of charging while

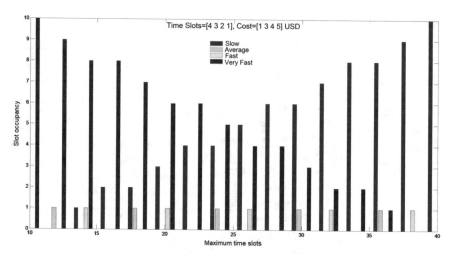

Fig. 4.3 Slot occupancy versus maximum time slot for different charging systems

satisfying all customers. The constraint C_1 shows that only one charging level can take place. C_2 ensures that the total charging time should be less than or equal to the total time T. C_3 indicates that variables representing the charging levels can be either 0 or 1. Assume that we have 10 EVs to be charged. The rates for different charging levels are as follows:

- For slow charging, 4 time slots costs 1 $
- For average charging, 3 time slots costs 3 $
- For fast charging, 2 time slots costs 4 $
- For very fast charging, 1 time slot costs 5 $

Solving the optimization problem yields the optimal scheduling of slow, average, fast, and very fast charging for EVs at different time slots as shown in Fig. 4.3. It is noticed that slot occupancy is high in the case when the maximum number of available slots is large. On the other hand, slot occupancy is high when the maximum number of available slots is small. This is because slow charging needs more slots to complete charging. The result shows the optimal number of slot occupancy for different charging types for the given number of maximum available slots.

4.5 Conclusion

The world's fossil fuel supply is diminishing rapidly, and the transportation sector is one of the major consumers. Further, to make our cities green and pollution-free, EVs must dominate the transportation sector in the future, and to successfully incorporate EVs into the intelligent transportation and IoT systems, optimal scheduling paradigms need to be developed to ensure optimal charging performance with lower prices. This chapter presented a scheduling approach to maximize total profit.

Chapter 5
Blockchain Technology for Security and Privacy in Internet of Things

5.1 Introduction

The Internet of Things (IoT) offers smart solutions by connecting physical objects through the Internet. The connectivity of IoT nodes can be wired and wireless. Nowadays, IoT nodes are involved in almost every walk of life with an objective to improve quality of life. For example, IoT-based e-health solutions are available and researchers in academia and industry are investigating more sophisticated solutions. Similarly, IoT-based solutions are available for intelligent transportation system (ITS), environment monitoring, etc. [19]. The number of IoT devices grows in number which generates a large amount of data over the Internet. The data generated by IoT devices is not only critical but also contains sensitive information. Many advanced communication technologies including cellular networks, ZigBee, Bluetooth, and cognitive radio networks are used to increase reliability and reduce delay in IoT systems. Recently, the third-generation partnership project (3GPP) standardized narrowband IoT (NB-IoT) to support IoT in long-term evaluation advanced (LTE-A) (as discussed in Chap. 2). The NB-IoT aims to provide low-rate connectivity for low-power IoT devices with extended coverage [86, 87]. The architecture of IoT mainly consists of four elements. This includes IoT nodes that can be a part of ad hoc sensor network, gateways which act as an intermediate device between IoT nodes and cloud infrastructure, cloud infrastructure, and application users as shown in Fig. 5.1.

The modern IoT solutions are being adopted rapidly which results in vulnerabilities, security risks, and cyber-attacks which should be investigated properly [88]. It is highlighted that "to the extent that everyday objects become information security risks, the IoT could distribute those risks far more widely than the Internet has to date [89]." IoT nodes are generally low power and do not have extensive computational capabilities. Further, IoT nodes are expected to perform main functionalities of applications. Therefore, traditional security protocols can be computationally expensive for low-power IoT devices. In addition, most of the

© The Author(s), under exclusive licence to Springer Nature Switzerland AG 2019
W. Ejaz and A. Anpalagan, *Internet of Things for Smart Cities*, SpringerBriefs in
Electrical and Computer Engineering, https://doi.org/10.1007/978-3-319-95037-2_5

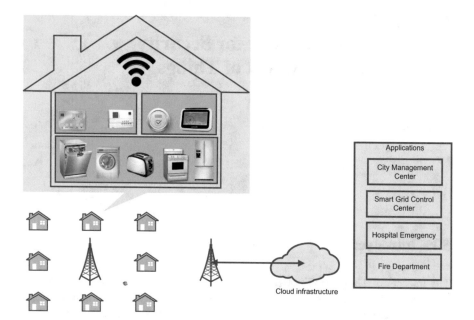

Fig. 5.1 IoT architecture

existing literature on cyber-attacks is designed for centralized networks which may not be a suitable option for highly centralized IoT systems [90]. Therefore, IoT systems require distributed, lightweight, and scalable solutions to prevent from the cyber-attacks.

The blockchain is the fundamental technology for the Bitcoin (the first cryptocurrency system) and is also considered as a candidate solution to address security issues, cyber-attacks, and privacy concerns in IoT systems [91]. The blockchain is a database that retains a large amount of data. For example, in case of bitcoin, the transactions of bitcoin are pushed into a block by users. The block is then amended with the blockchain once it is full. It uses a process called mining process in which some nodes attempt to solve a cryptographic puzzle which is a resource consuming job. Following the mining process, the new block is then appended with the blockchain. It is a distributed approach where each node in the chain has a copy of chain. The best thing about blockchain is that its scalable, nodes need to append to the chain. There are two main elements of blockchain technology: (1) the actions generated by users in the system (called transactions) and (2) transactions are recorded in blocks, where it is ensured that the blocks are in correct sequence and not altered. The blockchain technology has the capability to preserve the privacy of IoT users in a highly decentralized environment. Further, blockchain technology can help to specify and enforce different levels of access policies to restrict unauthorized operations on data generated by IoT devices. Authors in [92] discussed that the adoption of blockchain for IoT applications can be useful but not a straightforward process.

In this chapter, we provide a comprehensive survey of existing blockchain-based solutions for security and privacy in IoT systems. The objective is to provide a holistic view of blockchain technology for IoT systems. Following are the main contributions of this chapter:

- The existing literature was carefully analyzed to get a deeper understanding of research direction in the area of blockchain for IoT systems.
- We highlighted different challenges associated with the deployment of IoT and blockchain for the IoT systems.
- We present two case studies to investigate the performance of blockchain for IoT systems.
- We highlight the open research issues in blockchain for IoT systems.

Rest of the chapter is organized as follows: Sect. 5.2 provides an analysis of existing literature in the area of blockchain for IoT systems. Different challenges associated with the blockchain for IoT are highlighted in Sect. 5.3. Section 5.4 provides three case studies to investigate the performance of blockchain for IoT systems. Finally, the conclusion is drawn and open research issues in blockchain for IoT systems are discussed in Sect. 5.5.

5.2 Literature Review

Security and privacy issues in IoT emerged due to its global growth and continuous increase of data generated by IoT nodes. Recently, many researchers from academia and industry show the effectiveness of blockchain for IoT security and privacy. In [91], authors discussed decentralized approaches for security and privacy in IoT systems. The focus was given to the smart homes. Authors proposed a framework based on proposed modified blockchain for smart homes. The proposed scheme was analyzed in terms of basic security goals, i.e., confidentiality, integrity, and availability. Simulation results are presented to show the effectiveness of proposed modified blockchain scheme in the case of smart homes. A multi-layer secure network model based on blockchain is presented in [93]. The proposed model reduced the complexity and computation for the use of blockchain for IoT systems by dividing it into the multi-level decentralized network.

Authors in [94] investigated the prospect of blockchain for the information distribution in IoT systems. Key security requirements are highlighted and how blockchain can help to address these requirements are discussed. A design for information distribution in IoT systems using blockchain is presented to analyze that how existing security schemes can be made more powerful with the use of blockchain technology. In [95], authors highlighted several issues including integrity, anonymity, and adaptability for data management in IoT systems. Also, many use cases are discussed for the use of blockchain technology to address highlighted issues, as well as open research issues are pointed out to address the abovementioned issues in blockchain for IoT systems. A brief overview of blockchain for IoT is presented in [96]. Authors addressed how blockchain can address different challenges associated with IoT systems including costs and capacity constraint, deficient architecture, cloud server availability, and susceptibility to manipulation. Also, it is emphasized that how blockchain can improve the overall security in the IoT systems.

In [97], a food supply chain traceability system was designed based on hazard analysis and critical control points (HACCP), the blockchain, and IoT. The objective was to provide a platform for the members of supply chain securely, transparently, and reliably. Further, a new idea is presented for large-scale decentralized systems called BigchainDB. Also, challenges associated with future use of blockchain technology in advanced food supply chain traceability system are discussed. Authors in [98] proposed Internet of Smart Things (IoST) by adding features based on artificial intelligence. IoST uses a blockchain protocol (permission-based) called Multichain for secure communication among smart things. The choice of Multichain protocol in IoT systems was mainly because of its low communication cost.

A lightweight blockchain technology-based architecture for IoT is proposed in [99]. The proposed architecture can reduce the overhead of traditional blockchain schemes while providing the same level of security and privacy. Authors validated the proposed architecture in a smart home environment to highlight its effectiveness. It is demonstrated by simulation results that the proposed solution can significantly drop the packet and processing overhead when compared with traditional blockchain technology. In [100], authors studied the use of blockchain technology for better availability and accountability in IoT systems. An overview of the implementation of the ongoing prototype is provided for better understanding.

In summary, given in Table 5.1, blockchain technology is extensively investigated in the last couple of years. However, unlike traditional blockchain schemes, we need lightweight and computationally efficient schemes for incorporation in IoT systems.

5.3 Challenges Associated with Secure IoT Deployment and Blockchain for IoT

The new benefits offered by blockchain for IoT arrive with some new challenges. There are many challenges involved in both the secure deployment of IoT and blockchain for IoT. The objective of security and privacy of IoT systems are availability, integrity, and confidentiality similar to any communication systems. Some of the key challenges associated with the secure deployment of IoT systems include:

- IoT systems are highly fragmented and consist of a variety of protocols and communication technologies. This makes security and privacy issues very

Table 5.1 Summary of recent literature on Blockchain for IoT

Ref no.	Year	Objective	Solution	Remarks
[91]	2017	To provide a distributed approach for security and privacy for smart homes	Authors proposed a modified blockchain scheme for smart homes	The proposed scheme was analyzed in terms of basic security goals, i.e., confidentiality, integrity, and availability
[93]	2017	To reduce the complexity and computation for the use of blockchain for IoT systems	Authors divided IoT systems into multi-level decentralized network based on blockchain technology	The proposed multi-level network based on blockchain technology is a feasible solution for secure IoT network
[94]	2017	To investigate the prospect of blockchain for the information distribution in IoT systems	A design is presented to analyze that how existing security schemes can be made more power full with the use of blockchain technology	Authors discussed how Key security requirements can be satisfied by the use of blockchain technology
[95]	2016	To provide a systematic literature review on blockchain for the IoT	Many use cases are discussed for the use of blockchain technology to address highlighted issues as well as open research issues are pointed out in blockchain for IoT systems.	Three factors are taken into account, i.e., integrity, anonymity, and adaptability
[96]	2017	To check the feasibility of blockchain for IoT systems	Different challenges in IoT are highlighted and their potential solutions based on blockchain technology are presented	Overall, it is emphasized that how blockchain technology can improve security in IoT systems
[97]	2017	Design and development of food supply chain traceability system to provide a platform for the members of supply chain securely, transparently, and reliably	Authors proposed food supply chain traceability system based in HACCP, blockchain, and IoT	Challenges associated with future use of blockchain technology in advanced food supply chain traceability system are discussed
[98]	2017	Design and development of Internet of Smart Things (IoST) and use blockchain technology for secure communication	Authors used a permission-based blockchain protocol called Multichain for secure communication among smart things	Multichain protocol offers low communication cost and is a suitable choice for IoT solutions
[99]	2017	Develop a lightweight architecture based on blockchain technology for IoT systems	The proposed lightweight architecture was validated for the use case of smart homes	The proposed architecture offers less overhead in terms of packets and processing
[100]	2017	Study the effectiveness of blockchain for better availability and accountability in IoT systems	Develop a prototype of IoT system for better understanding	It is concluded that the availability is significantly improved using blockchain technology

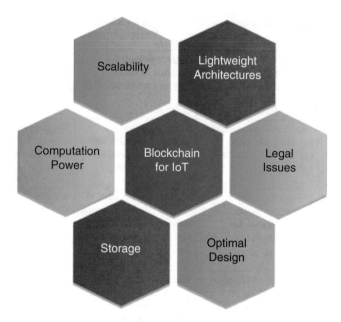

Fig. 5.2 IoT architecture

challenging. There is a need of standardization for better interoperability which will certainly reduce the complexity of IoT systems.

- IoT nodes in some applications are physically accessible which makes them prone to the physical attacks.
- Individual IoT nodes must be secured.
- IoT nodes can join and leave network according to their requirements. This needs a reliable lightweight authentication protocol for secure communication.
- A minimum level of security must be ensured for global deployment of IoT systems.
- We need to develop an International privacy standard for global deployment of IoT systems.

In addition to the challenges with the secure deployment of IoT systems, there are certain challenges associated with the use of blockchain technology for IoT systems as shown in Fig. 5.2.

- Scalability: It is important to test existing and design new blockchains for the scalable IoT systems.
- Lightweight architectures and schemes: Design and development of lightweight blockchain-based architectures for IoT systems is very important to reduce the overhead of traditional blockchains. However, the same level of security and privacy as traditional blockchains must be ensured.
- Computational Power: IoT systems are diverse with the wide range of capabilities. To perform encryption by all IoT nodes may not be possible in practical

scenarios. Therefore, some mechanisms should be devised to perform encryption using a group of IoT nodes or mechanism which has minimum overhead on IoT nodes.

- Storage: The blockchain technology is suitable for decentralized IoT systems because it lacks centralized controller. However, each IoT node needs to store the ledger which increases in size with the time. IoT nodes may not be capable to store a large amount of data.
- Optimal design: An optimal IoT system should be designed while considering blockchain-based security and privacy as a foundation element. This will result in an optimal design which gives equal precedence to connectivity, computation, coordination, security, and privacy.
- Legal Issues: The security and privacy standards vary in different countries and regions. This is a serious challenge for the successful adaptation of blockchain technology in IoT systems. There is need for standard framework that manufacturers can use for providing security and privacy solutions.

5.4 Case Studies

In this section, we will present two case studies which show the significance of blockchain technology in IoT systems.

5.4.1 Smart Homes

Smart home networks allow homeowners to use resources efficiently. A smart home can be equipped with a number of IoT nodes and sensors. Similar to the traditional IoT architecture, smart home architecture consists of: (1) sensors and devices, (2) communication network, and (3) cloud. In addition to the traditional elements of architecture, Blockchain-based architecture has local blockchain which is stored on a resource capable node. The resource capable node is called "miner" and is also responsible for communication between within and outside of the smart home. Further, a local storage is there to store ledgers of the blockchain. An analysis is performed to measure the performance of blockchain technology in smart homes. An overview of blockchain-based smart home architecture is presented in [91] as shown in Fig. 5.3. The proposed architecture offers less overhead for low-power IoT nodes. It also consumes less energy and less time for different transactions when compared with the traditional blockchain technology. However, the proposed architecture needs to be tested for other IoT applications. We should come up with an architecture which is suitable for many IoT applications.

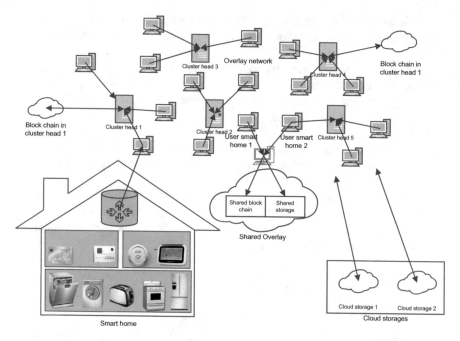

Fig. 5.3 Blockchain-based smart home architecture [91]

5.4.2 Food Supply Chain Traceability System

A traditional food supply chain consists of mainly five members: (1) production, (2) processing, (3) warehousing, (4) distribution, and (5) retail. Food traceability system is extremely important for the food safety. Authors in [97] presented a framework for food supply chain traceability system based on blockchain technology as shown in Fig. 5.4. It is a decentralized IoT system which uses sensors and communication technologies to collect and transfer data related to the food items. Each member mentioned above can add, update, and look at the information about the food item. Each food product is equipped with RFID tag which gives a unique identity. The members of this system also have the digital profile which contains information such as location, role in the supply chain, etc. The data is stored in a blockchain database which is accessible by each member. The members can register themselves in the system and after that, each member will have a public and private key. The proposed framework will provide real-time information about the safety of food products in a distributed way. The proposed system can significantly enhance the efficiency and transparency of the food supply chain. This will boost the confidence of end user in the food industry.

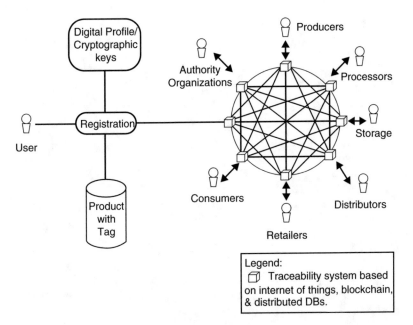

Fig. 5.4 Framework for food supply chain traceability system [97]

5.5   Conclusion

Security and privacy are prime issues for the success of IoT systems. In this chapter, we presented a comprehensive literature review of blockchain for IoT systems. We also outlined key challenges associated with the deployment of IoT and blockchain for IoT systems. We presented two case studies to illustrate the significance of blockchain for IoT. Despite extensive research on blockchain for IoT in the last couple of years, there exist several open areas which need to be investigated.

References

1. IBM's Smarter City Challenge. https://www.smartercitieschallenge.org/. Accessed 23 Oct 2017
2. Smart city success requires road maps, not free association. https://amsterdamsmartcity.com/posts/municipal-governments-overwhelmed-with-the-endless-choices-for-smart-city-projects-need-to-take-a-breath-and-plan-ahead. Accessed 23 Oct 2017
3. R. Giffinger, C. Fertner, H. Kramar, R. Kalasek, N. Pichler-Milanovic, E. Meijers, Smart cities. Ranking of European medium-sized cities, Final Report, Centre of Regional Science, Vienna UT, 2007
4. E. Mardacany, Smart cities characteristics: Importance of built environment components, 2014
5. L. Iliadis, H. Papadopoulos, C. Jayne, *Engineering Applications of Neural Networks Part 2* (Springer, Cham, 2013)
6. S. Parnell, S. Oldfield, *The Routledge Handbook on Cities of the Global South* (Routledge, Abingdon, 2014)
7. N. Kogan, K.J. Lee, Exploratory research on success factors and challenges of smart city projects. Asia Pac. J. Inf. Syst. **24**(2), 141–189 (2014)
8. Backgrounder: SmartTrack Stations & RER Implementation. https://www.toronto.ca/home/media-room/backgrounders-other-resources/backgrounder-smarttrack-stations-and-rer-implementation/. Accessed 01 May 2018.
9. A. Ahmed, M. Awais, M. Naeem, M. Iqbal, W. Ejaz, A. Anpalagan, H. Kim, Multiple power line outage detection in smart grids: a probabilistic Bayesian approach. IEEE Access **6**(1), 10650–10661 (2018)
10. M. Basharat, W. Ejaz, S.H. Ahmed, Securing cognitive radio enabled smart grid systems against cyber attacks, in *2015 First International Conference on Anti-Cybercrime (ICACC)* (IEEE, Piscataway, 2015), pp. 1–6
11. W. Ejaz, M. Naeem, A. Shahid, A. Anpalagan, M. Jo, Efficient energy management for the Internet of Things in smart cities. IEEE Commun. Mag. **55**(1), 84–91 (2017)
12. V.C. Gungor, D. Sahin, T. Kocak, S. Ergut, C. Buccella, C. Cecati, G.P. Hancke, Smart grid and smart homes: key players and pilot projects. IEEE Ind. Electron. Mag. **6**(4), 18–34 (2012)
13. A. Wheeler, Commercial applications of wireless sensor networks using ZigBee. IEEE Commun. Mag. **45**(4), 70–77 (2007)
14. N.A. Somani, Y. Patel, Zigbee: a low power wireless technology for industrial applications. Int. J. Control Theory Comput. Model. (IJCTCM) **2**(3), 27–33 (2012)
15. A. Boyano, P. Hernandez, O. Wolf, Energy demands and potential savings in European office buildings: case studies based on EnergyPlus simulations. Energy Build. **65**, 19–28 (2013)

16. B. Stiller, T. Bocek, F. Hecht, G. Machado, P. Racz, M. Waldburger, A new economic analysis of infrastructure investment, Department of Treasury with the Council of Economic Advisers. Technical report, March 2012

17. C. of Copenhagen, Copenhagen Intelligent Traffic Solutions. https://stateofgreen.com/en/profiles/city-of-copenhagen/solutions/copenhagen-intelligent-traffic-solutions. Accessed 20 Oct 2017

18. S. Pellicer, G. Santa, A.L. Bleda, R. Maestre, A.J. Jara, A.G. Skarmeta, A global perspective of smart cities: a survey, in *2013 Seventh International Conference on Innovative Mobile and Internet Services in Ubiquitous Computing (IMIS)* (IEEE, Piscataway, 2013), pp. 439–444

19. W. Ejaz, M. Ibnkahla, Machine-to-machine communications in cognitive cellular systems, in *2015 IEEE International Conference on Ubiquitous Wireless Broadband (ICUWB)* (IEEE, Piscataway, 2015), pp. 1–5.

20. M. Naeem, W. Ejaz, L. Karim, S.H. Ahmed, A. Anpalagan, M. Jo, H. Song, Distributed gateway selection for machine-to-machine communication in cognitive 5G networks. IEEE Netw. Mag. **31**(6), 94–100 (2017)

21. Z.-W. Alliance, The internet of things is powered by Z-Wave (2016). Retrieved January, vol. 28, p. 2016, 2016

22. A. Al-Fuqaha, M. Guizani, M. Mohammadi, M. Aledhari, M. Ayyash, Internet of things: a survey on enabling technologies, protocols, and applications. IEEE Commun. Surv. Tutor. **17**(4), 2347–2376 (FourthQuarter 2015)

23. S.S.I. Samuel, A review of connectivity challenges in IoT-smart home, in *3rd MEC International Conference on Big Data and Smart City (ICBDSC)* (IEEE, Piscataway, 2016), pp. 1–4

24. LinkLabs, A comprehensive look at low power, wide area neworks for IoT engineers and decision makers. https://www.link-labs.com/lpwan. Accessed 20 Oct 2017

25. S. Andreev, O. Galinina, A. Pyattaev, M. Gerasimenko, T. Tirronen, J. Torsner, J. Sachs, M. Dohler, Y. Koucheryavy, Understanding the IoT connectivity landscape: a contemporary m2m radio technology roadmap. IEEE Commun. Mag. **53**(9), 32–40 (2015)

26. V. Mohanan, R. Budiarto, I. Aldmour, *Powering the Internet of Things with 5G Networks* (IGI Global, Hershey, 2017)

27. LinkLabs, Selecting a wireless technology for new industrial Internet of Things Products. Technical report, 2015

28. A.W. Paper, What G is right for IoT/ M2M. Technical report, 2015

29. J. Swetina, G. Lu, P. Jacobs, F. Ennesser, J. Song, Toward a standardized common M2M service layer platform: Introduction to oneM2M. IEEE Wirel. Commun. **21**(3), 20–26 (2014)

30. F. Boccardi, R.W. Heath, A. Lozano, T.L. Marzetta, P. Popovski, Five disruptive technology directions for 5G. IEEE Commun. Mag. **52**(2), 74–80 (2014)

31. S.-G. Yoon, S.-G. Kang, S. Jeong, C. Nam, Priority inversion prevention scheme for PLC vehicle-to-grid communications under the hidden station problem. IEEE Trans. Smart Grid vol. PP, (99) (2017)

32. V.C. Gungor, D. Sahin, T. Kocak, S. Ergut, C. Buccella, C. Cecati, G.P. Hancke, Smart grid technologies: Communication technologies and standards. IEEE Trans. Ind. Inf. **7**(4), 529–539 (2011)

33. S. Ziegler, A. Skarmeta, P. Kirstein, L. Ladid, Evaluation and recommendations on IPv6 for the Internet of Things, in *2015 IEEE 2nd World Forum on Internet of Things (WF-IoT)*, Milan, December. (IEEE, Piscataway, 2015), pp. 548–552

34. M. Centenaro, L. Vangelista, A. Zanella, M. Zorzi, Long-range communications in unlicensed bands: the rising stars in the iot and smart city scenarios. IEEE Wirel. Commun. **23**(5), 60–67 (2016)

35. S. Jianjun, W. Xu, G. Jizhen, C. Yangzhou, The analysis of traffic control cyber-physical systems. Proc.-Soc. Behav. Sci. **96**, 2487–2496 (2013)

36. T. UslÃ et al., The trend towards the Internet of Things: what does it help in disaster and risk management? Planet@ Risk, vol. 3, no. 1, 2015

37. Ericsson, Mobility report: internet of things forecast. Technical report, 2018

38. M. Stolpe, The internet of things: Opportunities and challenges for distributed data analysis. ACM SIGKDD Explorations Newsletter **18**(1), 15–34 (2016)
39. Z.M. Hira, D.F. Gillies, A review of feature selection and feature extraction methods applied on microarray data. Adv. Bioinform. **2015**, 198363 (2015)
40. H.H. Pajouh, R. Javidan, R. Khayami, D. Ali, K.-K.R. Choo, A two-layer dimension reduction and two-tier classification model for anomaly-based intrusion detection in IoT backbone networks, in *IEEE Transactions on Emerging Topics in Computing*, 2016
41. T. Zhang, B. Yang, Big data dimension reduction using PCA, in *2016 IEEE International Conference on Smart Cloud (SmartCloud)* (IEEE, Piscataway, 2016), pp. 152–157
42. K. Guo, Y. Tang, P. Zhang, CSF: crowdsourcing semantic fusion for heterogeneous media big data in the Internet of Things. Inf. Fusion **37**, 77–85 (2017)
43. M.H. ur Rehman, V. Chang, A. Batool, T.Y. Wah, Big data reduction framework for value creation in sustainable enterprises. Int. J. Inf. Manage. **36**(6), 917–928 (2016)
44. S. He, D.-H. Shin, J. Zhang, J. Chen, Y. Sun, Full-view area coverage in camera sensor networks: dimension reduction and near-optimal solutions. IEEE Trans. Veh. Technol. **65**(9), 7448–7461 (2016)
45. A. Papageorgiou, B. Cheng, E. Kovacs, Real-time data reduction at the network edge of Internet-of-Things systems, in *11th International Conference on Network and Service Management (CNSM)* (IEEE, Piscataway, 2015), pp. 284–291
46. S. Cheng, Z. Cai, J. Li, H. Gao, Extracting Kernel dataset from big sensory data in wireless sensor networks. IEEE Trans. Knowl. Data Eng. **29**(4), 813–827 (2017)
47. Dimensionality Reduction. https://www.ritcheng.com/machine-learning-dimensionality-reduction/. Accessed 22 July 2018
48. Beginners Guide To Learn Dimension Reduction Techniques. https://www.analyticsvidhya.com/blog/2015/07/dimension-reduction-methods/. Accessed 22 July 2018
49. C.P. Chen, C.-Y. Zhang, Data-intensive applications, challenges, techniques and technologies: a survey on Big Data. Inf. Sci. **275**, 314–347 (2014)
50. F. Chen, P. Deng, J. Wan, D. Zhang, A.V. Vasilakos, X. Rong, Data mining for the internet of things: literature review and challenges. Int. J. Distrib. Sens. Netw. **11**(8), 431047 (2015)
51. D.H. Jeong, C. Ziemkiewicz, W. Ribarsky, R. Chang, C.V. Center, Understanding principal component analysis using a visual analytics tool, Charlotte visualization center, UNC Charlotte, vol. 19, 2009
52. N. Kumar, S. Singh, A. Kumar, Random permutation principal component analysis for cancelable biometric recognition. Appl. Intell. **48**(9) 2824–2836 (2018)
53. J.C. Faria, C.G.B. Demétrio, I.B. Allaman, Biplot of multivariate data based on principal components, 2018
54. R. Indhumathi, S. Sathiyabama, Reducing and clustering high dimensional data through principal component analysis. Int. J. Comput. Appl. **11**(8), 1–4 (2010)
55. A. Asuncion, D. Newman, UCI machine learning repository, 2007
56. P. Sadeghi-Barzani, A. Rajabi-Ghahnavieh, H. Kazemi-Karegar, Optimal fast charging station placing and sizing. Appl. Energy **125**, 289–299 (2014)
57. A. Zakariazadeh, S. Jadid, P. Siano, Multi-objective scheduling of electric vehicles in smart distribution system. Energy Convers. Manage. **79**, 43–53 (2014)
58. P. Yi et al., Energy scheduling and allocation in electric vehicle energy distribution networks, in *Innovative Smart Grid Technologies (ISGT) IEEE PES*, February 2014, pp. 1–6
59. A. Al-Fuqaha, M. Guizani, M. Mohammadi, M. Aledhari, M. Ayyash, Internet of things: a survey on enabling technologies, protocols, and applications. IEEE Commun. Surv. Tutor. **17**(4), 2347–2376 (FourthQuarter 2015)
60. N. Taheri, *Linear Optimization Methods for Vehicle Energy and Communication Networks* (Stanford University, Stanford, 2012)
61. W. Su, M.Y. Chow, Computational intelligence-based energy management for a large-scale PHEV/PEV enabled municipal parking Deck. Appl. Energy **96**, 171–182 (2012)

62. M.C. Falvoa, R.L. Lamedicaa, R. Bartonib, G. Maranzanoc, Energy management in metro-transit systems: an innovative proposal toward an integrated and sustainable urban mobility system including plug-in electric vehicles. Electr. Power Syst. Res. **81**, 2127–2138 (2011)
63. J. Yu, W. Gu, Z. Wu, Intelligent PHEV charging and discharging strategy in smart grid, in *IEEE Fifth International Conference on Advanced Computational Intelligence(ICACI)*, October 2012, pp. 1107–1112
64. S. Ahmed, V. Ganesh, Plug-in vehicles and renewable energy sources for cost and emission reductions. IEEE Trans. Ind. Electron. **58**, 1229–1238 (2011)
65. T. David, B. Ross, The evolution of plug-in electric vehicle-grid interactions. IEEE Trans. Smart Grid **3**, 500–505 (2012)
66. A. Grünewald, S. Hardt, M. Mielke, and R. Brück, A decentralized charge management for electric vehicles using a genetic algorithm, in *IEEE World Congress on Computational Intelligence WCCI*, June 2012, pp. 1–7
67. R.S. Wimalendra, L. Udawatta, S. Karunarathna, Determination of maximum possible fuel economy of HEV for known drive cycle: genetic algorithm based approach, in *International Conference on Information and Automation for Sustainability*, December 2008, pp. 289–294
68. B. Özden, Modeling and optimization of hybrid electric vehicles. The Graduate School of Natural and Applied Sciences of Middle East Technical University, 2013
69. A. Bandyopadhyay, L. Wang, V.K. Devabhaktuni, R.C. Green, Aggregator analysis for efficient day-time charging of plug-in hybrid electric vehicles, in *IEEE Power and Energy Society General Meeting*, July 2011, pp. 1–8
70. Y. Yao, D.W. Gao, Charging load from large-scale plug-in hybrid electric vehicles: impact and optimization, in *IEEE PES Innovative Smart Grid Technologies (ISGT)*, February 2013, pp. 1–6
71. F. Fazelpour, M. Vafaeipour, O. Rahbari, M.A. Rosen, Intelligent optimization of charge allocation for plug-in hybrid electric vehicles utilizing renewable energy considering grid characteristics, in *IEEE International Conference on Smart Energy Grid Engineering*, August 2013, pp. 1–8
72. J. Prasomthong, W. Ongsakul, J. Meyer, Optimal placement of vehicle-to-grid charging station in distribution system using particle swarm optimization with time varying acceleration coefficient, in *International Conference and Utility Exhibition on Green Energy for Sustainable Development (ICUE)*, March 2014, pp. 1–6
73. D. Dallinger, J. Link, M. Büttner, Smart grid agent: plug-in electric vehicle. IEEE Trans. Sustainable Energy **5**(3), 710–717 (2014)
74. T. Ghanbarzadeh, P.T. Baboli, M. Rostami, M.P. Moghaddam, M.K. Sheikh-El-Eslami, Wind farm power management by high penetration of PHEV, in *IEEE Power and Energy Society General Meeting*, July 2011, pp. 1–5
75. A. Mercurio, A.D. Giorgio, F. Purificato, Optimal fully electric vehicle load balancing with an ADMM algorithm in Smartgrids, in *Mediterranean Conference on Control and Automation (MED)*, June 2013, pp. 119–129
76. A. Hess et al., Optimal deployment of charging stations for electric vehicular networks, in *International Conference on Emerging Networking Experiments and Technologies*, December 2012, pp. 1–6
77. D. Nguyen, L. Le, Optimal energy trading for building microgrid with electric vehicles and renewable energy resources, in *Innovative Smart Grid Technologies Conference (ISGT)*, 2014 February, pp. 1–5
78. M. Alizadeh, A. Scaglione, J. Davies, K. Kurani, A scalable stochastic model for the electricity demand of electric and plug-in hybrid vehicles. IEEE Trans. Smart Grid **5**(2), 848–860 (2014)
79. Q. Guo, S. Xin, H. Sun, Z. Li, B. Zhang, Rapid-charging navigation of electric vehicles based on real-time power systems and traffic data. IEEE Trans. Smart Grid **5**(4), 1969–1979 (2014)
80. C. Battistelli, L. Baringob, A. Conejob, Optimal energy management of small electric energy systems including V2G facilities and renewable energy sources. Electr. Power Syst. Res. **92**, 50–59 (2012)

81. M. Moradijoz, M.P. Moghaddam, Optimum allocation of parking lots in distribution systems for loss reduction, in *IEEE Power and Energy Society General Meeting*, July 2012, pp. 1–5

82. D.Q. Xu, G. Joós, M. Lévesque, M. Maier, Integrated V2G, G2V, and renewable energy sources coordination over a converged fiber-wireless broadband access network. IEEE Trans. Smart Grid **4**(3), 1381–1390 (2013)

83. A.U. Haq, M.J. Chadhry, F. Saleemi, A smart charging station for EVs with evaluation of different energy storage technologies, in *IEEE Conference on Clean Energy and Technology (CEAT)*, November 2013, pp. 248–253

84. N.G. Omran, S. Filizadeh, Location-based forecasting of vehicular charging load on the distribution system. IEEE Trans. Smart Grid **5**(2), 632–641 (2014)

85. M.R. Aghaebrahimi, M.M. Ghasemipour, A. Sedghi, Probabilistic optimal placement of EV parking considering different operation strategies, in *IEEE Mediterranean Electrotechnical Conference*, April 2014, pp. 108–114

86. A.E. Mostafa, Y. Zhou, V.W. Wong, Connectivity maximization for narrowband IoT systems with NOMA, in *2017 IEEE International Conference on Communications (ICC)* (IEEE, Piscataway, 2017), pp. 1–6

87. Y.-P.E. Wang, X. Lin, A. Adhikary, A. Grovlen, Y. Sui, Y. Blankenship, J. Bergman, H.S. Razaghi, A primer on 3GPP narrowband internet of things. IEEE Commun. Mag. **55**(3), 117–123 (2017)

88. S. Sicari, A. Rizzardi, L.A. Grieco, A. Coen-Porisini, Security, privacy and trust in internet of things: the road ahead. Comput. Netw. **76**, 146–164 (2015)

89. A. Bassi, G. Horn, Internet of things in 2020: a roadmap for the future. Eur. Commission: Inf. Soc. Media **22**, 97–114 (2008)

90. R. Roman, J. Zhou, J. Lopez, On the features and challenges of security and privacy in distributed internet of things. Comput. Netw. **57**(10), 2266–2279 (2013)

91. A. Dorri, S.S. Kanhere, R. Jurdak, P. Gauravaram, Blockchain for IoT security and privacy: the case study of a smart home, in *2017 IEEE International Conference on Pervasive Computing and Communications Workshops (PerCom Workshops)*. (IEEE, Piscataway, 2017), pp. 618–623

92. A. Dorri, S.S. Kanhere, R. Jurdak, Blockchain in internet of things: challenges and solutions (2016). Preprint. arXiv:1608.05187

93. C. Li, L.-J. Zhang, A blockchain based new secure multi-layer network model for internet of things, in *2017 IEEE International Congress on Internet of Things (ICIOT)* (IEEE, Piscataway, 2017), pp. 33–41

94. G.C. Polyzos, N. Fotiou, Blockchain-assisted information distribution for the internet of things, in *2017 IEEE International Conference on Information Reuse and Integration (IRI)* (IEEE, Piscataway, 2017), pp. 75–78

95. M. Conoscenti, A. Vetro, J.C. De Martin, Blockchain for the internet of things: a systematic literature review, in *2016 IEEE/ACS 13th International Conference of Computer Systems and Applications (AICCSA)* (IEEE, Piscataway, 2016), pp. 1–6

96. N. Kshetri, Can blockchain strengthen the internet of things? IT Professional **19**(4), 68–72 (2017)

97. F. Tian, A supply chain traceability system for food safety based on HACCP, blockchain & internet of things, in *2017 International Conference on Service Systems and Service Management (ICSSSM)* (IEEE, Piscataway, 2017), pp. 1–6

98. N. Fabiano, The internet of things ecosystem: the blockchain and privacy issues. The challenge for a global privacy standard, in *2017 International Conference on Internet of Things for the Global Community (IoTGC)* (IEEE, Piscataway, 2017), pp. 1–7

99. M. Samaniego, R. Deters, Internet of smart things-IoST: using blockchain and clips to make things autonomous, in *2017 IEEE International Conference on Cognitive Computing (ICCC)* (IEEE, Piscataway, 2017), pp. 9–16

100. A. Dorri, S.S. Kanhere, R. Jurdak, Towards an optimized blockchain for IoT, in *Proceedings of the Second International Conference on Internet-of-Things Design and Implementation* (ACM, New York, 2017), pp. 173–178

Index

Printed in the United States
By Bookmasters